In Situ Hybridization:

Medical Applications

In Situ Hybridization: Medical Applications

Edited by

G.R. Coulton and J. de Belleroche
Department of Biochemistry, Charing Cross and Westminster Medical School, London, UK

SPRINGER SCIENCE+BUSINESS MEDIA, B.V.

British Library Cataloguing in Publication Data

In situ hybridization.
I. Coulton, G.R. II. De Belleroche, J.
612.8042

Library of Congress Cataloging-in-Publication Data

In situ hybridization: medical applications / edited by Gary R. Coulton and Jackie de Belleroche.
p. cm.
Includes bibliographical references and index.
ISBN 978-94-010-5326-6 ISBN 978-94-011-2984-8 (eBook)
DOI 10.1007/978-94-011-2984-8
1. In situ hybridization. I. Coulton, Gary R. II. De Belleroche, J., 1945–
[DNLM: 1. Gene Expression. 2. Histocytochemistry. 3. Nucleic Acid Hybridization. 4. Nucleic Acids—analysis. 5. Nucleic Acids—diagnostic use. QU 58 s623]
QH452.8.I53 1992
616.07'58—dc20
DNLM/DLC
for Library of Congress 91-46508
CIP

Lasertypeset by Martin Lister Publishing Services, Bolton-le-Sands, Carnforth, Lancs.

Contents

List of Contributors

John McCafferty[1] and Carl Alldus[2]
[1] Cambridge Antibody Technology, Daly Laboratories, Babraham, Cambridge, UK
[2] Glaxo Research Limited, Greenford Road, Greenford, Middlesex, UK

Giorgio Terenghi and Julia M. Polak
Histochemistry Department, Royal Postgraduate Medical School, Hammersmith Hospital, Du Cane Road, London, UK

Jackie de Belleroche, Lisa Virgo, Amat Rashid and Yolanda Collaço Moraes
Department of Biochemistry, Charing Cross and Westminster Medical School, Fulham Palace Road, London, UK

Adrienne L. Morey and Kenneth A. Fleming
University of Oxford, Nuffield Department of Pathology and Bacteriology, John Radcliffe Hospital, Headington, Oxford, UK

Ton K. Raap[1] and C.J. Cornelisse[2]
[1] Department of Cytochemistry and Cytometry, Medical Faculty, Leiden University, Wassenaarseweg 72, 2333 AL Leiden, The Netherlands
[2] Department of Pathology, Medical Faculty, Leiden University, Wassenaarseweg 72, 2333 AL Leiden, The Netherlands

Preface

The molecular mechanisms that control cellular processes such as proliferation, growth and differentiation have continually fascinated researchers. Recently, enormous advances have been made in our understanding of the genome and messenger RNA expression as well as technological advances in areas such as nucleic acid sequencing and the polymerase chain reaction (PCR). One technique which is making an increasingly important contribution is *in situ* hybridization which is the subject of this book.

***In situ* hybridization** has developed as a method for localising specific DNA and RNA sequences. The great strength of this approach is that it allows the distribution of specific nucleic acids to be related to the protein products of the target gene by means of immunohistochemistry and each of these with particular cellular structures. Complimentary DNA, RNA or oligonucleotide probes, suitably labelled, are hybridized to specific DNA or RNA targets within tissues. The spatial information thus obtained has contributed greatly to our understanding of such diverse areas of research as gene mapping, viral infection, cytogenetics, gene expression, pre-natal diagnosis and development.

This book is not intended as just another collection of recipes, although it does describe theoretical and practical aspects of the technology. Rather, we critically describe the contribution made by *in situ* hybridization to specific areas of medical research.

McCafferty and Alldus describe the basis of *in situ* hybridization, providing just the right degree of theoretical background in order to understand how the method may be manipulated to the advantage of the investigator.

The great investigative strength of combining *in situ* hybridization with immunohistochemistry is described in the chapter by Terenghi and Polak. This approach allows us to compare the site and degree of expression of a particular messenger RNA with the distribution of the translated protein. Thus a more complete functional description of a particular cell type can be obtained.

This theme is extended by de Belleroche *et al.* in their description of messenger RNA detection in the nervous system. We see here that *in situ* hybridization is not simply a qualitative, descriptive tool but, with appropriate experimental technique, lends itself to quantitation.

One of the areas where *in situ* hybridization is firmly established as a technique of choice is in the localisation and diagnosis of viral disease. The advent of the polymerase chain reaction appeared to overtake the role of diagnosis from *in situ* hybridization. However, the mere presence of a known nucleic acid does not tell you anything about the cellular distribution, which in many instances will tell us more about the intracellular activity of the virus and hence is in turn is of greater prognostic value. In their chapter Morey and Fleming describe their use of *in situ* hybridization for localization of viral RNA and DNA in biopsies of human tissues.

The field of cytogenetics is currently undergoing a renaissance. *In situ* hybridization has for many years been very successfully employed for gene mapping purposes on metaphase spreads. Indeed it has been referred to as "the method of choice for gene mapping". However, metaphase spreads are very difficult to prepare from some types of tissues and it would be useful to be able to localise specific genes in interphase cells. Raap and Cornelisse describe the exquisite application of non-radioisotopic *in situ* hybridization to the cytogenetic investigation of the interphase nucleus. Of particular interest here is the ability to produce multiple labelling of gene loci. This approach is appropriate when using so-called "chromosome painting probes" as well as probes to specific single-copy genes.

We do not appear to have reached the limit of technical and biological innovation for *in situ* hybridization. The current trend towards the technical refinement of non-radioisotopic labelling systems will continue and expand the range of biological systems which lend themselves to this type of investigation seems almost endless. *In situ* hybridization is the next link in the chain once a particular genetic locus is described, and there are an awful lot of new genes out there to be discovered . . .

Colour Plates

Plate 1A & 1B Pituitary sections of (A) control and (B) thyroidectomized rat hybridized with ^{35}S-labelled β-TSH probe, and immunostained for β-TSH peptide ($\times$800). Note the presence of immunostaining and hybridization signal in the same cells. Haematoxylin counterstain.

Plate 2 Induction of c-fos-mRNA in cerebral cortex by unilateral injection of excitotoxin. Distribution of c-fos mRNA after unilateral injection of kainate into nucleus basalis to show a large increase in c-fos mRNA in cerebral cortex ipsilateral to the lesion (right hand side) relative to the contralateral unlesioned cerebral cortex (left hand side). The result coincides with that obtained by Northern blot analysis (Wood and de Belleroche, 1990) but allows localization of the signal to specific regions and cell types.

Plate 3 Distribution of Go α subunit mRNA in rat brain. Thin sections of rat brain were hybridized with Go α subunit cDNA. High concentrations of Go α subunit mRNA were found in cerebral cortex, hippocampus and cerebellum.

Plate 4 Triple colour detection of three different targets in a parvovirus-infected erythroid precursor. Red cell membrane protein glycophorin-a was labelled by the APAAP technique (red), host cell chromatin by a digoxigenin-labelled probe for whole human DNA (anti-digoxigenin peroxidase conjugate with DAB substrate; brown) and parvovirus DNA by a biotinylated probe (avidin alkaline-phosphatase conjugate with NBT-BCIP substrate; purple).

Plate 5 Combined APAAP labelling of macrophages with monoclonal antibody KP-1 (Fast Red substrate) and *in situ* hybridization for parvovirus B19 DNA using a digoxigenin labelled probe (NBT-BCIP purple-black substrate). A macrophage containing several parvovirus-infected nuclei is seen in the centre.

Plate 6 Triple fluorescence *in situ* hybridization with Bluescript plasmid DNA libraries from flow-sorted chromosomes 2, 4 and 8. The DNAs were labelled by nick-translation using, respectively, digoxigenin-, biotin- and fluorescein-dUTP. After a preannealing step with total human DNA, to compete out dispersely occurring repeat sequences, the DNAs were *in situ* hybridized to a normal metaphase plate, after which immunocytochemical detection followed with, respectively AMCA, TRITC and FITC as the microscopic reporter molecules. No counterstaining, triple exposure.

Plate 7 Triple fluorescence *in situ* hybridization to an interphase cancer cell using satellite DNAs for chromosome 1 (blue), 6 (green) and 17 (red). Labelling of the DNAs was as described in plate 6. The fluorescein signal was not amplified immunocytochemically. Note the numerical aberrations, No counterstaining, triple exposure.

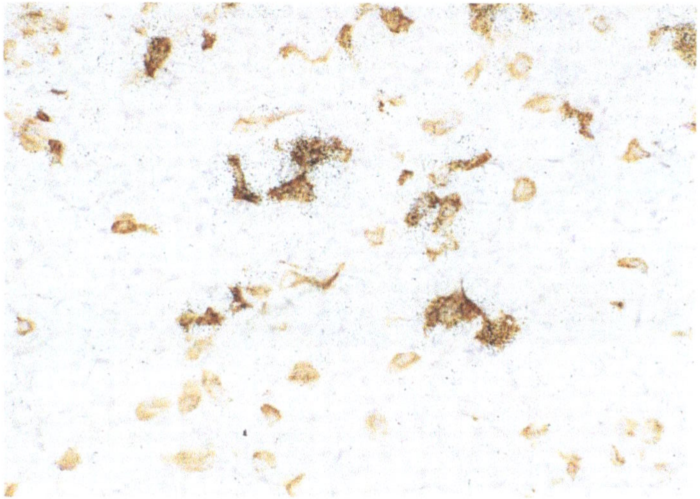

A

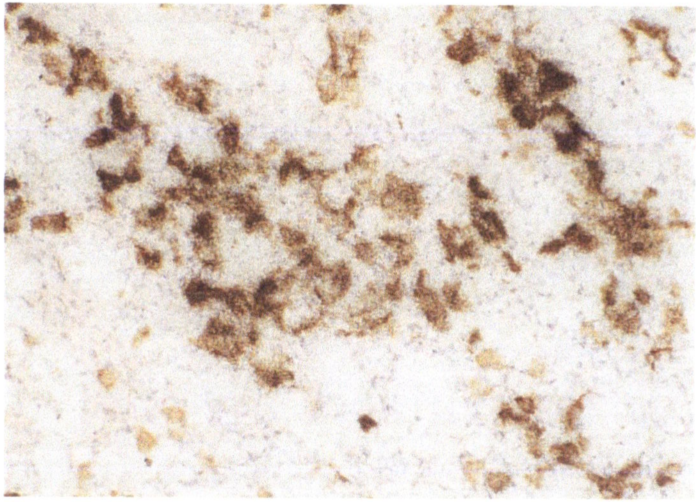

1B

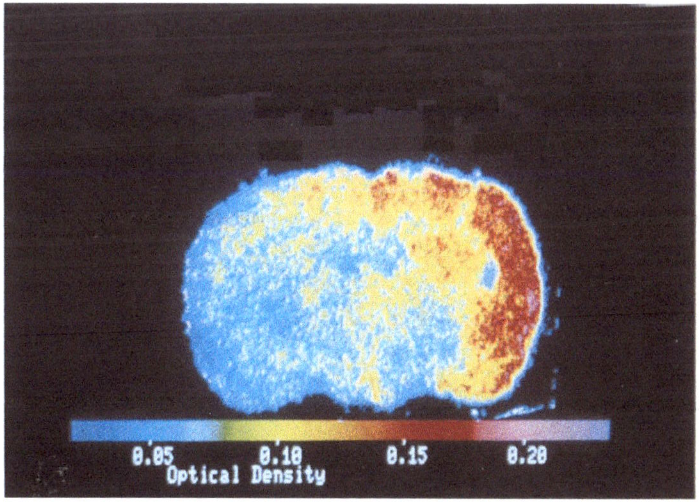
0.05 0.10 0.15 0.20
Optical Density

2

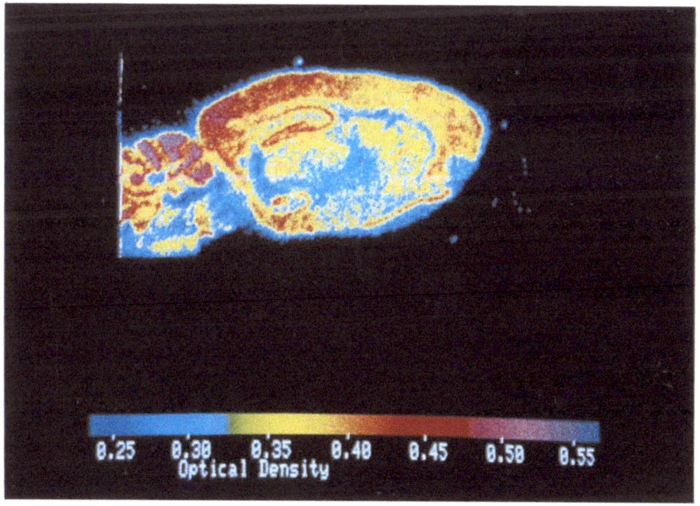
0.25 0.30 0.35 0.40 0.45 0.50 0.55
Optical Density

3

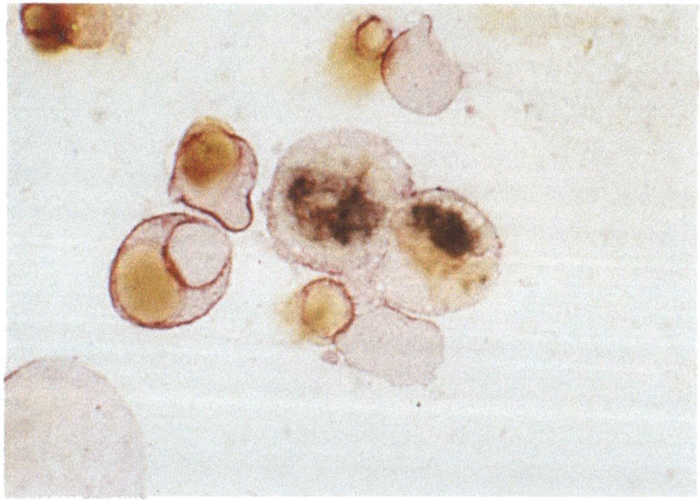

4

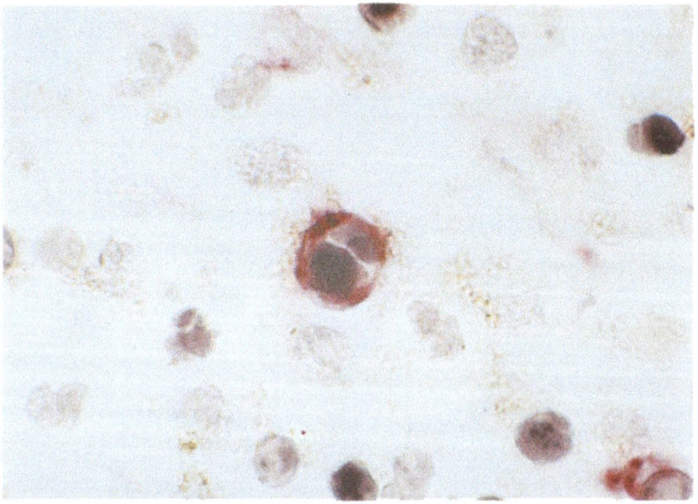

5

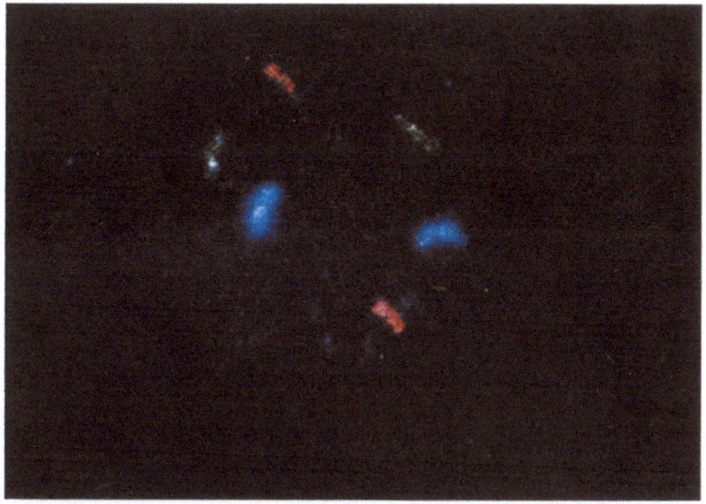

6

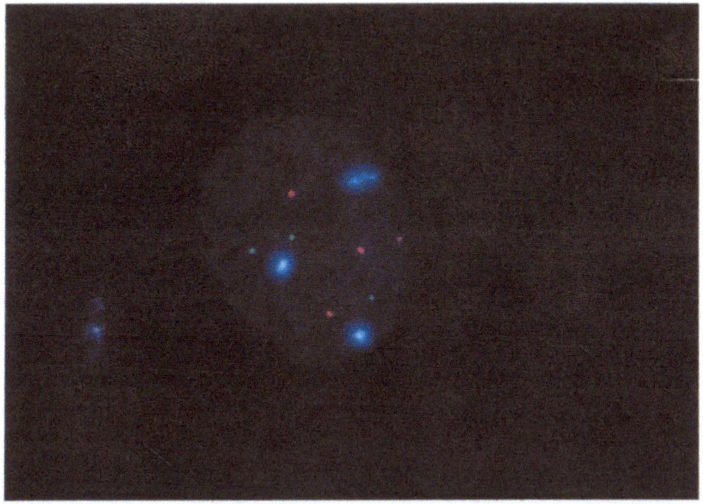

7

1
Techniques for *in situ* hybridization histochemistry

John McCafferty[1] and Carl Alldus[2]
[1]*Cambridge Antibody Technology, Daly Laboratories, Babraham, Cambridge, UK*
[2]*Glaxo Research Limited, Greenford Road, Greenford, Middlesex, UK*

1. INTRODUCTION

The range of approaches taken by different laboratories to carry out *in situ* hybridization can at first appear bewildering. The situation is further complicated by apparently contradictory recommendations from different authors. It is the aim of this chapter to provide a guide through this methodological "maze" and, where possible, to make recommendations on the choice of various parameters. It is not intended to provide complete protocols for the steps involved (see McCafferty *et al.*, 1989; Singer *et al.*, 1986).

In situ hybridization can be envisaged as having three main steps:

1. Tissue fixation and permeabilization.
2. Hybridization and stringency washing.
3. Detection.

The aim of tissue fixation is to preserve tissue morphology and to access the nucleic acids therein for hybridization. Additional permeabilization steps are often included to aid in accessing target.

Hybridization describes the process whereby the bases of a single stranded nucleic acid form hydrogen bonds with bases on a nucleic acid of a complementary base sequence to produce a relatively stable double stranded hybrid molecule. While there are a few additional factors to be considered for *in situ* hybridization, in principle the basic rules governing hybridization during *in situ* hybridization are the same as those affecting filter bound nucleic acids. After hybridization, unhybridized probe is removed by washing. The ability to distinguish specific signal from non-specific signal is of particular importance when using *in situ* hybridization and background must be kept to a minimum especially when high sensitivity is required. Thus adequate washing after hybridization is essential.

While recognizing these distinctive stages it should be noted that all the steps involved are intimately linked and variations at one stage may affect other stages. For example the choice of fixative may affect the need for pre-treatments. Conversely the pre-treatments used in a given protocol may "determine" the optimal fixative. Alternatively, the combination of fixative/pre-treatments used may affect subsequent probe diffusion. This, in turn, will affect the time needed for complete hybridization at a given probe concentration. The interdependence of such variables may, in part, explain apparently conflicting reports in the literature. In addition different demands are imposed on the system depending on whether cellular mRNA or DNA is being visualized. Target DNA in most cases will be nuclear, double-stranded and bound by histone and other chromosomal proteins. In the majority of cases, target RNA is cytoplasmic, and may be associated with ribosomes or ribonucleoproteins. Target RNA is also susceptible to digestion by RNAse enzymes and care must be taken to avoid this. The discussion that follows will assume RNA as the target but will highlight where differences occur with DNA as target.

2. TISSUE PREPARATION

The procedures described in this section are critical to the success of any *in situ* hybridization experiment, and separate *in situ* hybridization from other nucleic acid hybridization techniques. The protocols relating to tissue preparation are probably more variable than for any other stage of the *in situ* hybridization procedure. This section should clarify why these differences exist, and provide guidance as to which protocols suit the reader's particular needs.

2.1 Tissue handling

With all tissue handling procedures one should aim to maintain the tissue or cell morphology while minimizing nucleic acid degradation. The latter is of utmost importance when RNA is the target of choice. This is because RNases are robust and ubiquitous enzymes which will be found to varying degrees within the tissues and cells under manipulation. As a consequence, cellular RNA may exhibit a short half-life once cells have been harvested or tissues excised (10 minutes has been quoted for c-myc mRNA: Hoefler, 1987).

Furthermore, degradation of RNA may be accelerated by the release of lysosomal contents during handling of tissue (Johnson *et al.*, 1986). For this reason, cells and tissues should be fixed as quickly as possible. It is difficult to give precise guide-lines for the maximum time permitted between harvest and fixation but generally 30 minutes should not be exceeded (Hoefler *et al.*, 1986).

Some workers fix tissue sections on slides after they have been cut from frozen tissue blocks (Haase *et al.*, 1984). This has the advantage that RNA or DNA can be isolated from the same tissue blocks for analysis on blots (Hoefler, 1987), but it is not recommended, as it prolongs the delay before fixation and can compromise morphology.

We would recommend that the tissue be cut into suitable blocks (about 1 cm^3), and immersed immediately into the appropriate fixative. The tissue should be manipulated under nuclease free conditions (see section 2.3). While this is a convenient and rapid method, problems occasionally arise from limited diffusion of fixative. Tournier *et al.* (1987) have shown using rat liver that there is heterogeneity of *in situ* hybridization signal through tissue blocks fixed by immersion. This may be circumvented by cutting smaller blocks or by extending immersion. If this is not possible perfusion fixation may be adopted (Haase *et al.,* 1984; Moench *et al.*, 1985). Here the blood system is perfused with fixative, such that the tissue of interest is fixed prior to excision. Consequently no manipulation of unfixed specimens is necessary. This method is of particular value with tissues, such as the brain, where fixative penetration is poor.

If the tissue is to be frozen, then tissue blocks should be cryoprotected in 15% sucrose after fixation. This prevents the occurrence of tissue artefacts resulting from crystallization of water within the block. The cryoprotected tissue can then be stored in buffer at 4°C for up to two months, but it is advisable to prepare frozen cryostat blocks as soon as possible. Freezing should be performed rapidly to avoid formation of water crystals. Once frozen, the tissue blocks may be stored at –70°C or in liquid nitrogen for a much longer period of time before sectioning.

Cryostat sections, typically 4–10 μm thick, are collected onto RNase free, glass microscope slides and air dried. During the *in situ* hybridization procedure, the tissue sections will be exposed to quite harsh treatments, and therefore loss of tissue from the slides can sometimes be a problem (Moench *et al.*, 1985). To overcome this, a variety of slide treatments may be employed to promote better adhesion of the tissue. Treatments commonly used include gelatin-chrome alum (Hoefler, 1987), 0.5% gelatin (Lawrence and Singer, 1985), triethoxysilylpropylamine (TESPA) (Berger, 1986) and 3 × Denhardts (Moench *et al.*, 1985). Our studies, however, have shown poly-L-lysine (PLL) of molecular weight >150 000 to be most effective over a range of fixatives.

Paraffin embedded sections have been successfully used in *in situ* hybridization experiments (Haase *et al.*, 1984; Lum, 1986; Tourner *et al.*, 1987). This facilitates retrospective studies of archival specimens (Burns *et al.*, 1988; Morley and Hodes, 1987). However, it should be noted that paraffin embedding can cause tissue shrinkage, which may produce problems with delicate tissues. As an alternative, Bandtlow *et al.* (1987) successfully performed *in situ* hybridization on delicate iris tissue embedded in 11% gelatin.

When adherent cultured cells are under investigation, the most simple method of attaching them to a matrix is to grow them directly on microscope slides or coverslips (Singer and Ward, 1982; McCafferty *et al.*, 1989). The slides or coverslips should be placed in petri-dishes containing the appropriate cell suspension. Alternatively, multiwell slides may be used, in which case droplets of cell suspension are placed onto the wells. The latter is particularly useful, as different cell lines may be grown on different wells of the same multiwell slide.

The slides or coverslips should be RNase free and sterile, the latter being achieved by immersion in 70% ethanol. In our experience, no treatments to enhance cell adhesion are necessary (McCafferty *et al.*, 1989), as the cells will bind quite adequately. Cells grown directly onto slides have normal morphology and a well spread cytoplasm, which improves visualization of signal when hybridizing to mRNA (Cumming and Fallon, 1988). They can also be transferred from cell culture medium to fixative within seconds, reducing the likelihood of target degradation.

If the cells of interest grow in suspension, or the above procedure is not feasible, then cytocentrifugation can be employed for attachment of cells to slides. The cells are harvested and then centrifuged at an appropriate concentration (about 10^5 cells/ml) onto PLL treated slides. The cells should be fixed as soon as possible after cytocentrifugation. Cytocentrifugation is a physically harsh process which may disrupt cell morphology. As a result, some investigators have sedimented cells passively onto slides (Berger, 1986; Lawrence *et al.*, 1988). This method, however, may result in poor cell adhesion and should only be considered when using "sticky" cells such as lymphocytes.

Regardless of the method of attachment, after fixation the cells should finally be dried by baking at 37°C for at least 4 hours. This results in improved tissue adhesion. Both cells and tissue sections can be stored in sealed boxes at –70°C for several months. The boxes should contain activated silica gel to absorb any moisture.

2.2 Choice of fixative

A variety of different fixatives are recommended within the literature on *in situ* hybridization, making selection of the most appropriate reagent a confusing task to anyone new to the technique. In order to try to rationalize the choice, it is wise to consider the aims of fixation and the types of fixative available. The fixation step should maintain cell or tissue morphology and retain the target nucleic acid in a form accessible to the probe. To achieve this, there are two classes of fixative available: cross-linking and precipitating.

Cross-linking fixatives (e.g. paraformaldehyde and glutaraldehyde) are alkylating agents which cross-link nucleic acids and proteins to form a network within the cell. As such, they are effective at maintaining tissue morphology and retaining target sequences. However, excessive cross-linking may result in a matrix that hinders probe diffusion and impairs annealing of bases during hybridization. This can be overcome by including permeabilization steps in the tissue pretreatment (section 2.3), or by using shorter hybridization probes (Moench *et al.*, 1985).

In contrast, target nucleic acids in cells or tissues fixed with a precipitating agent (e.g. ethanol/acetic acid or methanol/acetone) are highly accessible to probe, and subsequent permeabilization steps are rarely necessary. However, there are frequently problems with tissue morphology and target retention using these fixatives (Moench *et al.*, 1985, Singer *et al.*, 1986). Lawrence and Singer (1985) have demonstrated that tissues fixed with ethanol/acetic acid can lose up to 75% of their target RNA during the *in situ* hybridization procedure.

While there are suggestions that different tissues have different optimal fixatives, 4% paraformaldehyde is an excellent starting point that will suffice in most cases. It is probably the most commonly used fixative (Berger, 1986; Hoefler *et al.* 1986; Hoefler, 1987; Larsson *et al.*, 1988; Terenghi *et al.*, 1987; McCafferty *et al.*, 1989), being a cross-linking agent which still permits accessibility of probe to target, in some cases without the need for additional permeabilization steps (Lawrence and Singer, 1985; McCafferty *et al.*, 1989; Singer *et al.*, 1986). Due to its excellent protein-retaining properties, paraformaldehyde is the most suitable fixative for use when performing *in situ* hybridization and immunocytochemistry on the same sections (Shivers *et al.*, 1986). It also produces a low intrinsic autofluorescence (Singer *et al.*, 1986), which is important if fluorescent detection is to be employed.

The time of fixation is important, as over-fixation can reduce probe accessibility (Bandlow *et al.*, 1987; Brigati *et al.*, 1983; Singer and Ward, 1982). The optimal time will depend on the permeability of the tissue to the fixative, and will be shorter for cells and monolayers than for intact tissue. With paraformaldehyde we use 30 minutes for cells and 60 minutes for tissues (at room temperature).

If satisfactory results cannot be obtained with paraformaldehyde, even with

	Fixative	*Fixation time*
	Cells	Tissues
2% Glutaraldehyde	30 mins, RT	60 mins, RT
Bouins Fixative 75% picric acid, 25% formalin, 1% acetic acid	30 Mins, RT	60 mins, RT
Ethanol: acetic acid (95:5)	15 mins, RT	30 mins, RT
Methanol:acetone	4 mins, –20°C	20 mins, –20°C

RT – room temperature

subsequent permeabilization steps, then we suggest that the following fixation regimes are tested:

When *in situ* hybridization studies are carried out on retrospective samples such as paraffin embedded tissues, the fixative is pre-determined. In such cases, the tissue pretreatments (section 2.3) have to be optimized to give the best possible results with the fixative used. Indeed, the interdependence of fixation and tissue pretreatments cannot be overstressed, and this should always be remembered when choosing protocols.

2.3 Tissue pretreatment

There are a number of possible treatments that may be applied to tissues or cells prior to hybridization to increase *in situ* hybridization signal, reduce background or provide experimental controls (Guitteny *et al.*,1988; Moench *et al.*, 1985, Singer *et al.*, 1986). These are discussed below, and advice is given as to when each procedure should be employed. The pretreatment protocol finally adopted will depend on other stages of the *in situ* hybridization procedure, particularly fixation, but as a general rule the number of pretreatment steps should be kept to a minimum. This is because they not only increase the time and complexity of an experiment, but can also cause a deterioration in tissue morphology and leaching of target nucleic acids.

When RNA is used as probe or target, it is essential that the pretreatments are performed under RNase-free conditions. Despite fixation of the cells or tissue, contaminating RNases may still degrade the target RNA. Furthermore, they may remain in the cell and degrade RNA probes during hybridization. To prevent RNase contamination solutions should be prepared with double-distilled water and sterilized by autoclaving. Alternatively they may be treated with 0.1% diethyl-pyrocarbonate (DEPC) at 37°C for about 12 hours (Maniatis *et al.*, 1982). Glassware and slide racks should be baked at 250°C for at least 4 hours or treated with DEPC. Gloves should also be worn throughout (Maniatis *et al.*, 1982).

Prior to pretreatment, paraffin-embedded sections should be dewaxed, and all cells and tissues rehydrated.

2.3.1 Enzymatic

In cases where a strongly cross-linking fixative has been used, proteolytic treatment of the tissue may be required. This will permeabilize the cell, permitting probe entry, and remove proteins associated with the target that might otherwise inhibit hybridization. Proteases most commonly used are pronase (Berger, 1986), proteinase K (Singer and Ward, 1982), and pepsin (Burns *et al.*, 1988). Proteolytic treatment should be carried out in nuclease-free buffer at the appropriate temperature. Proteinase K can be obtained nuclease-free, but pronase should be pre-digested by incubation at 37°C for about 4 hours to remove any contaminating nucleases (Berger, 1986). The protease concentration and time of incubation should be titrated for each type of tissue used. Proteolytic treatment is not normally necessary when a precipitative fixative has been used.

If the target nucleic acid is DNA, then cross hybridization of the probe to RNA should be prevented by treating the tissue with RNase prior to hybridization. The RNase should be rendered DNase-free by boiling stocks before use. Typically, the tissue is incubated in a 100μg/ml solution of RNase A at 37°C for 30 minutes. Thorough washing and a fixation step are necessary at the conclusion of RNase treatment in order to prevent continued activity of the enzyme. This is particularly important if an RNA probe is to be used. As an alternative, Lawrence *et al.* (1988) carried out this step after hybridization using RNase H.

Target nucleic acid may intentionally be degraded using RNase or DNase to provide negative controls (section 7). Once again, the nuclease digestion should be followed by thorough washing and fixation to prevent continued enzyme activity.

2.3.2 Non-enzymatic

Detergents may be used to increase tissue permeability. Triton X-100, a non-ionic detergent which solubilizes membrane proteins, is usually favoured, and is used at a concentration from 0.01% to 0.3% (Hoefler *et al.*, 1986; McCafferty *et al.*, 1989).

Tissues and cells may be subjected to dilute acid treatment, typically 0.2 M hydrochloric acid, followed by a high temperature wash. This denatures basic proteins, and can be used in conjunction with proteolytic treatment to facilitate their removal. This not only results in a more accessible target, but because basic proteins will attract nucleic acid probes, it can also reduce background (Brahic and Haase, 1984; Haase *et al.*, 1984).

2.3.3 Treatments to reduce backgrounds

The sugar/phosphate "backbone" of nucleic acid probes is negatively charged, with the result that probes often bind to positive charges on both the tissue and the microscope slide. This non-specific binding can result in high backgrounds, and is most often a problem when large and/or biotinylated probes are used (Lawrence and Singer, 1985). To counteract this, specimens can be treated with 0.25% acetic anhydride in 0.1 M triethanolamine pH 8 (Cox *et al.*, 1984; Hoefler *et al.*, 1986; Singer and Ward, 1982). This reagent neutralizes positive charges on both the tissue and the microscope slide by acetylation of amino groups (Hayashi *et al.*, 1978). For the treatment to work effectively, the rack of slides should be immersed in the triethanolamine solution prior to addition and mixing of acetic anhydride.

Measures may also be taken to reduce non-isotopic detection system background. For example, slides may be immersed in periodic acid and hydrogen peroxide to block phosphatases or peroxidases if these enzymes form part of the detection system (section 6.2; Cogg *et al.*, 1986).

2.3.4 Prehybridization and DNA denaturation

Some protocols recommend a prehybridization step (Bandtlow *et al.*, 1987) which is usually carried out at the same temperature as hybridization, and in an identical buffer (section 5.2) but with the exclusion of the probe. This means that the tissue becomes equilibrated with components of the hybridization buffer, and "sticky" sites therein are blocked by components such as sodium pyrophosphate and unlabelled RNA or genomic DNA (Brahic and Haase, 1978). Despite these potential advantages, we, along with other investigators, have found that prehybridization has no beneficial effects.

If the target nucleic acid is DNA, then it must be denatured prior to hybridization. Hybridization buffer containing the probe is added to the tissue section or cells as described in section 5. The slides are then heated to between 90°C and 100°C for about ten minutes, denaturing both the DNA target and the probe (McDougall *et al.*, 1988). The slides are then cooled to the appropriate temperature for hybridization.

In summary, pretreatments are used to facilitate hybridization of probe with target (by both enzymatic and non-enzymatic means) and to reduce background. A typical pretreatment regime used in our lab (derived from Hoefler *et al.*, 1986) is as follows: 0.3% Triton X-100 in phosphate buffered saline (PBS) for 15 minutes, PBS washes (2 × 3 minutes), digestion with pronase (1–10 μg/ml) at 37°C (30 minutes), immersion in 0.1 M glycine/PBS (5 minutes), post-fixation in paraformaldehyde (3 minutes) and blocking with 0.25% acetic anhydride in 0.1 M triethanolamine (McCafferty *et al.*, 1989).

In practice we find pretreatments are not always necessary (McCafferty *et al.,* 1989). In fact Singer *et al.* (1986; 1986) have found their pretreatment regime to be detrimental to the final signal.

3. CHOICE OF PROBE

Over the years a range of nucleic acid probe types have been used for *in situ* hybridization. In general double stranded cloned DNA probes have been utilized most (Busch *et al.*, 1987; Godard, 1983; Hafen *et al.*, 1983; Venezky *et al.*, 1981). Hu and Messing (1982) introduced the use of single stranded DNA probes derived from sequences cloned into M13 vector. In addition single stranded chemically synthesized oligonucleotide probes have found increasing use with the greater availability of DNA synthesis technology. In recent years there has been increased use of RNA probes and these offer several advantages. Their increased use has been facilitated by the availability of cloning vectors containing bacteriophage-specific RNA polymerase promoters which permit transcription from cloned sequences. Labelling systems involving many of the above probe types are commercially available in kit formats.

3.1 Double stranded DNA probes

3.1.1 Nick translation

Nick translation (Rigby *et al.*, 1977; Amersham, 1987) makes use of the double action of *E. coli* DNA polymerase as a means of introducing labelled nucleotides into DNA probes. A 5′→ 3′ exonuclease activity initiates at nicks in the DNA and removes nucleotides from the nicked strand. Simultaneously, a 5′→ 3′ DNA polymerase activity fills in this strand using labelled nucleotide triphosphates, resulting in the nick being "translated" along the strand.

The nicks in the template are introduced by DNase I and the size of probe generated is dependent on the concentration of this enzyme. Apart from affecting size, greater numbers of nicks access more template for DNA polymerase and the rate of labelling is increased. Nick translation reactions are generally carried out from 30 minutes to 5 hours at 15°C. This lower temperature is used to prevent overall degradation of template by the combined activities of DNase I and the exonuclease activity of DNA polymerase.

The specific activity of probe achieved using this technique is dependent on both the specific activity of the nucleotide triphosphate used and on the degree of replacement of nucleotides in the DNA substrate. The degree of replacement increases with time to a point which depends on the relative amount of template and nucleotide. If one wishes to maximize replacement (and therefore specific activity) then the unincorporated nucleotides should not limit the reaction but should be in molar excess over the nucleotides of the template. Thus for 1 μg of template containing 750 pmoles of each nucleotide (assuming equivalent concentrations of each nucleotide in the sequence) one should use at least 1000 pmoles of the labelled nucleotide precursor (unlabelled nucleotides are usually present in much greater excess). Specific activity of the probe can be increased further using a second labelled precursor although the reaction rate may not be as high with 2 precursors at relatively low concentration.

Using nick translation it is possible to achieve homogeneous labelling to a high specific activity (10^8–10^9 dpm/μg) of microgram amounts of DNA in a relatively short time (1–2 hours). In addition, genomic, plasmid or phage DNA can be used without any further manipulations or sub-cloning.

3.1.2 Random primer labelling

As an alternative to nick translation random primer labelling may be used (Feinberg and Vogelstein, 1983, 1984; Amersham, 1987). This method utilizes single stranded nucleic acid templates but double stranded DNA may be used after denaturation. The template is allowed to hybridize to short oligonucleotide sequences. Random hexamer sequences are commonly used for this purpose. DNA polymerase then extends from this double stranded region to produce a labelled double stranded molecule. The Klenow fragment of DNA polymerase, which lacks the 5′–3′ exonuclease activity described earlier, gives better results in this protocol and permits greater flexibility with respect to time and temperature of reaction.

Another advantage of random priming over nick translation is that relatively impure DNA, such as samples contaminated with agarose, can be labelled. Thus it is possible to separate DNA fragments in low melting agarose cut out the fragment of interest, and label this sequence in isolation of other sequences such as those of the vector which might contribute to background. Using random priming, probes of high specific activity (greater than 10^9 dpm/μg) can be generated. It must be remembered however that the original template is still present thereby reducing the effective specific activity for double stranded DNA probes.

Higher than expected signals have been reported for large (>1.5 kB) double stranded probes (Lawrence and Singer, 1985) and this has been attributed to the formation of "networks" of probe. If a probe sequence adjacent to vector sequences hybridizes to target then this will leave part of the molecule unhybridized, namely the vector sequences. This single stranded sequence can in turn act as a target for complementary vector sequences.

More single stranded regions will be generated forming a network of probe and resulting in an amplification of signal at the position of the original target. The benefit of this amplification has to be balanced against the potential increase in background caused by vector sequences and with the variability in signal found when this type of amplification occurs.

3.2 Single stranded DNA probes

3.2.1 M13 derived probes

Single stranded probes have an advantage over double stranded probes in that there is no re-annealing of probe and thus a higher effective probe concentration is maintained throughout the hybridization. In fact single stranded DNA probes from bacteriophage M13 vectors have been reported to give greater sensitivity than double stranded probe when used for *in situ* hybridization (Goedert, 1986).

The M13 bacteriophage passes through one phase in its life cycle, where its DNA is single stranded and one where it is double stranded. The double stranded form can be cleaved with restriction enzymes and used to clone sequences of interest. One strand of the inserted sequence will subsequently pass through a single stranded phase on re-introduction to a bacterial host. The orientation of the insert with respect to the vector will determine which insert strand becomes single stranded.

Probes can be generated by hybridizing a primer sequence to the single stranded M13 DNA, upstream of the inserted sequence, and elongating through the inserted sequence using the Klenow fragment of DNA polymerase (Berger, 1986; Hu and Messing, 1982). The size of probe generated may be controlled by choosing an appropriate concentration of nucleotide precursors such that the labelling of vector sequences is minimized. Alternatively the reaction may be chased with an excess of unlabelled nucleotides. The larger double stranded product can then be cleaved with a restriction enzyme. After denaturation and electrophoresis the short single stranded insert sequence is separated from the rest of the DNA. This approach is a little cumbersome but may be of use in labs which already have inserts of interest in M13 vectors.

3.2.2 Synthetic oligonucleotide probes

The increased availability of DNA synthesizers has enabled researchers to obtain oligonucleotide probes from any known sequence and has to an extent reduced the need to receive such probes from fellow workers. In addition, this obviates the need for the skill and materials required for plasmid and phage DNA preparations. Oligonucleotide probes of 25–50 nucleotides have been used successfully for *in situ* hybridization (Berger, 1986; Lewis *et al.*, 1985;

Lewis *et al.*, 1986; Lynch *et al.*, 1987; Uhl *et al.*, 1986). While access of probe to target is facilitated using these short probes, one major disadvantage of oligonucleotide probes is that their shorter size results in a much reduced target coverage and therefore reduced sensitivity.

Oligonucleotide probes can be labelled in a number of ways. 5′ end labelling involves the addition of a single labelled phosphate group to the 5′ end of a nucleic acid molecule. In a reaction catalysed by T4 polynucleotide kinase, the labelled γ-phosphate group of ATP is transferred to the 5′ hydroxyl group at the 5′ end of the oligonucleotide (Amersham, 1987; Maniatis *et al.*, 1982). This 5′ hydroxyl is available for reaction in synthetic oligonucleotides but is protected with 5′ phosphate groups in DNA and RNA. These can be removed using alkaline phosphatase (Maniatis *et al.*, 1982). While this approach provides a convenient method for labelling oligonucleotides the main drawback is that only a single label is added to each probe molecule (Lewis *et al.*, 1985; Lynch *et al.*, 1987).

3′ terminal labelling utilizes the ability of terminal deoxynucleotidyl transferase to add on nucleotides to the 3′ end of single stranded DNA. This allows addition of several labelled nucleotides and has been used to detect rat arginine vasopressin and pro-opiomelanocortin (Lewis *et al.*, 1985).

Finally, oligonucleotides with a number of labels may be generated by hybridizing a short (e.g. 10 mer) oligonucleotide to a longer one (e.g. 50 mer and extending it using the Klenow fragment of DNA polymerase and labelled nucleotide(s). The labelled and unlabelled molecules may be separated by electrophoresis if required to generate a single stranded probe of defined sequence with multiple labels (Berger, 1986; Uhl *et al.*, 1986).

3.3 RNA probes

In recent years there has been greater use of RNA probes for *in situ* hybridization (Constantini *et al.*, 1988; Cox *et al.*, 1984; Hoefler *et al.*, 1986; Lynch *et al.*, 1987). RNA probes can be easily generated when the sequence of interest is cloned into a vector containing a bacteriophage RNA polymerase promoter upstream of the insert (Green *et al.*, 1983; Mifflin *et al.*, 1987). These promoters act as very specific recognition sequences for bacteriophage RNA polymerases (Butler and Chamberlin, 1982; Chamberlin *et al.*, 1983). Thus the SP6 bacteriophage of *Salmonella typhimurium* produces an RNA polymerase which specifically recognizes a promoter sequence from this bacteriophage. Polymerase/promoter pairs from *E. coli* bacteriophages T3 and T7 are also commonly used in this way.

Many of the vectors such as pSP64 used for producing RNA probes have a polylinker cloning sequence downstream of an RNA polymerase promoter.

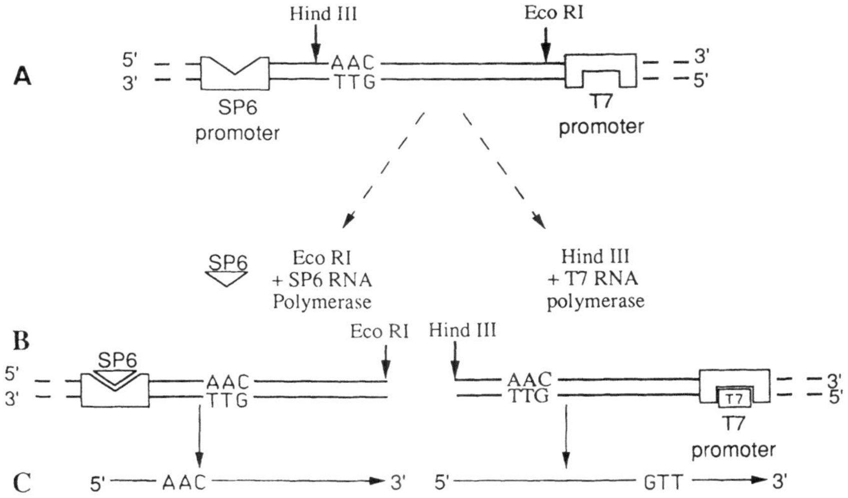

Figure 1 Generation of RNA probes. **A** represents a dual promoter vector for generating RNA probes. Three base pairs from one end of the insert are highlighted to follow the outcome. The dotted line represents plasmid sequence and the boxes represent the two promoters for the two different RNA polymerases. **B** represents the result of cleaving the template with a suitable restriction enzyme and addition of the accompanying RNA polymerase. **C** shows the final RNA transcripts generated from both enzymes.

This provides a convenient site for introduction of an insert. DNA prepared from clones generated in this way will produce single stranded RNA probes when mixed with ribonucleotide triphosphates, the appropriate enzyme and buffer. In order to prevent a read through of vector sequences, which can contribute to background in subsequent hybridizations, the template is usually cleaved immediately downstream of the inserted sequence. The polylinker can often provide a useful cleavage site. This then results in an RNA transcript whose size is defined by the distance between the initiating nucleotide and the cleavage site. The usefulness of this approach has been improved further in pAM18, Bluescribe and other vectors where the polylinker is flanked on both sides by two different RNA polymerase promoters.

Transcripts can be generated from either strand of the cloned sequence using templates cleaved on either side of the insert in combination with the appropriate RNA polymerase (Figure 1).

If a transcript is generated with the same sequence as cellular mRNA this is designated as being in the sense orientation and will not hybridize to the cellular mRNA. Transcription from the opposite strand will give rise to a non-identical but complementary strand which will hybridize. This is termed an anti-sense transcript (Figures 1 and 2).

The purification of RNA probes is relatively simple and involves a short DNase I step to remove template followed by a precipitation step. In a standard labelling reaction using a limiting concentration of ^{32}P-labelled UTP at 12.5 μM, incorporation of 70–90% of label is common and yields 115–150 ng from 1 μg of DNA template. Using an excess of all nucleotides (500 μM each), incorporation rates of 10–50% are achieved and yield 0.7–3 μg of RNA. Incorporation of UTP substituted with biotin is usually less efficient (1–30% incorporation).

Large RNA probes may be hydrolysed to a smaller size to improve penetration into cells. It is possible to hydrolyse in a controlled way by incubating with 40 mM $NaHCO_3$, 60 mM Na_2CO_3 (pH 10.2) at 60°C. The time required for hydrolysis is calculated from the formula:

$$t = \frac{L_o \, L_f}{kL_o - L_f}$$

where t is the time required in minutes, L_o is the original length, L_f is the final average fragment length required (in kilobases) and k is the rate constant. Using the conditions described above the value of k is approximately 0.11 kb/minute. Cox *et al.* (1984) hydrolyse down to an average size of 0.15 kb and in our experience 0.1–0.25 kb is best (McCafferty *et al.*, 1989).

RNA probes have a number of advantageous features. When carrying out an *in situ* hybridization experiment the specificity of signal achieved needs to be checked (sections 6 and 7). One way of doing this is to use a probe which does not hybridize to the tissue under examination. As stated earlier RNA probes provide the facility for generating sense or anti-sense probes. Sense probes provide a useful negative control in the detection of mRNA since they would not be expected to hybridize in most cases and have the same melting temperature (section 5) as the anti-sense probe (Figure 2).

RNA probes are single stranded and as such are not prone to the problem of re-annealing discussed earlier for double stranded probes. Improved sensitivity has been demonstrated on blots (Melton *et al.*, 1984) and by *in situ* hybridization (Bloch *et al.*, 1986; Cox *et al.*, 1984). Another advantage is the increased stability of duplexes involving RNA. This allows higher washing temperatures to be used and may reduce non-specific background.

Perhaps a greater contribution to reducing background is the ability to use RNase A in one of the post-hybridization washes. This enzyme digests single stranded but not double stranded RNA molecules. Thus non-specifically bound probe will be digested, but hybrid molecules remain intact. While it is possible to distinguish signal from background using Northern and Southern

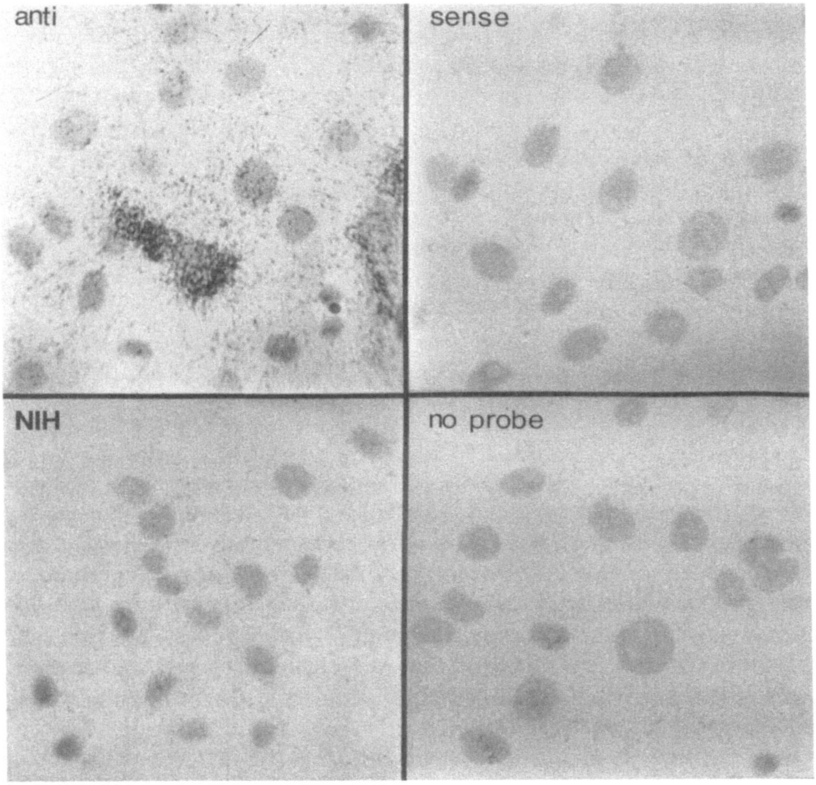

Figure 2 *In situ* hybridization using RNA probes: *In situ* hybridization was done on T15 cells (derived from mouse NIH 3T3 cells) transfected with the human N-ras gene) which had been grown onto slides and fixed in 4% paraformaldehyde. ^{32}P labelled human N-ras RNA probes were prepared from a vector with dual RNA polymerase promoters as depicted in Figure 1 and these probes were hydrolysed to an average size of 250 bases. The cells were hybridized for 1 hour at a probe concentration of 500 ng/ml. No pretreatments were carried out in this instance. The top row shows cells probed with anti-sense or sense RNA. The bottom row shows hybridization to the parental NIH 3T3 cell line which has not been transfected with human N-ras (left) and T15 cells which have been hybridized in the absence of probe (see McCafferty *et al.*, 1989).

blotting, this is not possible with a single *in situ* sample. *In situ* hybridization indicates only the presence or absence of signal and anything which reduces background will therefore ultimately lead to greater sensitivity.

In summary, single stranded probes appear to have greater sensitivity than double stranded probes although the latter has been used successfully for *in situ* hybridization. Among the single stranded probes, perhaps the easiest way to generate a new probe within the laboratory is to make an oligonucleotide (provided one has access to a DNA synthesizer). The problem is that the small target size covered by oligonucleotide probes means a loss of sensitivity. Our preference is for RNA probes. RNA probes cover a larger target area and can be hydrolysed down to improve access to target within the cell, the facility for generating sense transcripts provides a useful negative control in most cases and the ability to wash at high temperatures and with RNase A leads to a low background. This means that greater sensitivity can be achieved and that higher concentrations of probe can be used to increase the rate of probe diffusion/hybridization.

M13 probes also cover larger target areas than oligonucleotides but the preparation of probe is rather involved.

4. CHOICE OF LABEL

Several types of radioactive and non-radioactive labels have been used to measure hybridization of probe to tissue by *in situ* hybridization. Radioactively labelled probes are detected by dipping slides in photographic emulsion and developing the slides after an appropriate exposure time to produce silver grains near or over the target cells. One problem with radioactive probes is the additional safety and containment aspects that have to be considered when handling radioactivity (further discussion of this aspect is beyond the scope of this chapter).

Reduced stability is also a problem with some radioactive probes. Apart from radioactive decay the probe itself is subject to radiolysis. The higher the specific activity the greater the problem. ^{32}P labelled probes with specific activities of 10^3–10^9 dpm/μg may be used for 10–15 days when stored at –20°C (Amersham, 1987). It is not advisable to exceed this duration. Non-radioactive labels do not suffer from the same problems of safety or stability. Originally they were not found to be as sensitive as radioactive probes but in recent years detection levels approaching that of ^{32}P probes have been reported (Heiles *et al.*, 1988, Hopman *et al.*, 1986; Jablonski *et al.*, 1986; Mifflin *et al.*, 1987; Renz and Kurz, 1984). In addition, non-radioactive probes have the potential for higher resolution than radioactive probes since signal is generated at the site of probe localization and not at distant sites in an emulsion.

4.1 Radioactive labels

Radioisotopes commonly used for labelling nucleic acids are ^3H (tritium), ^{32}P, ^{35}S and ^{125}I (Berger, 1986; Cox *et al.*, 1984; Gall and Pardue, 1969; Godard,

Table 1 Characteristics of nucleotides used in nucleic acid labelling

Radionucleotide	*Half-life*	*Type of emission*	*Maximum energy of emission* (MeV)	*Specfic activity range* (Ci/mmol)
^{32}P	14.3 days	β	1.71	400–6000
^{35}S	87.4 days	β	0.167	400–1500
^{125}I	60 days	γ	0.035	1000–2000
		β	0.035	
^{3}H	12.43 years	β	0.018	25–100

(Taken from ref.3)

1983; Hafen *et al.*, 1983; Hoefler *et al.*, 1986; Lynch *et al.*, 1987; Mifflin *et al.*, 1987; Syrjänen *et al.*, 1988). Each of these has its own characteristics with respect to half-life, maximum energy of emission (MeV) and path length in emulsion (Table 1) which in turn affect stability of probe and resolution and sensitivity of signal.

The main reason for using *in situ* hybridization is to localize nucleic acids to a single cellular, sub-cellular or chromosomal location and so the best isotope for achieving highest resolution is tritium with a path length of 1μm in emulsion (MeV reflects the energy of the particle which in turn determines its path length in the emulsion). The main problem with tritium however is its poor sensitivity. Exposure times of several months are not uncommon (Figure 3). This is largely due to the low specific activities achievable with this isotope.

^{35}S with a path length of 15–20μm is less useful for determining chromosomal localization but is useful for distinguishing which cells express a particular mRNA. It does have the advantage that results are achieved more quickly than tritium. Use of ^{35}S nucleotides may result in high background due to non-specific sticking of nucleotides but this can be overcome using high concentrations of dithiothreitol, β-mercaptoethanol or α-thioUTP in the hybridization buffer (Cox *et al.*, 1984; Syrjänen *et al.*, 1988).

The most rapid results can be achieved using ^{32}P labelled probes. The high energy of the β particle however means that path lengths of several hundred microns are possible and so resolution suffers. While this is true, the majority of grains formed are much closer to the target cell and a degree of resolution to within 1 or 2 cell diameters is possible (Figure 3). Apart from the reduced exposure time required using ^{32}P labelled probes, one also has the facility to make a more rapid assessment of an *in situ* hybridization experiment by exposing slides to X-ray film in advance of dipping. Some laboratories use ^{32}P labelled probes to optimize conditions for *in situ* hybridization or to rapidly screen tissues/probes and then, when necessary, use tritium or ^{35}S labelled probes for higher resolution results.

^{125}I is less commonly used (Hayashi *et al.*, 1978; Allen *et al.*, 1987) because

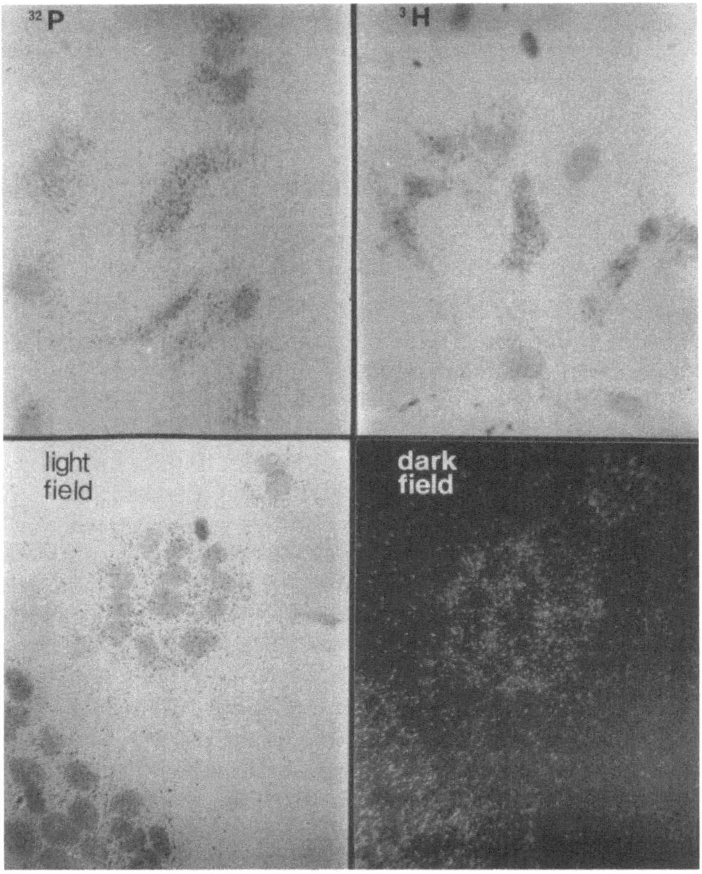

Figure 3 Comparison of tritium with ^{32}P labelled probes and dark field/light field microscopy: *In situ* hybridization was carried out on N872 cells (derived from mouse NIH 3T3 cells transfected with the human N-ras gene (from Chris Marshall, (ICR)). These were treated as described for Figure 2. Anti-sense probes were labelled either with ^{32}P (top left) or tritium (top right). Human A431 cells (see McCafferty et al., 1989) were grown onto glass slides and treated as described in Figure 2. The same field was viewed either in light field (bottom left or dark field (bottom right).

of problems with high background and the extra safety precautions required when dealing with the penetrating γ radiation from this source.

4.2 Non-radioactive labels

4.2.1 Biotin

The most commonly used non-radioactive probe system utilizes the interaction between biotin (vitamin H) and the protein avidin. This interaction has a very high affinity constant ($K_A = 10^{15}\ M^{-1}$) and so a very strong non-covalent interaction is observed between these two molecules.

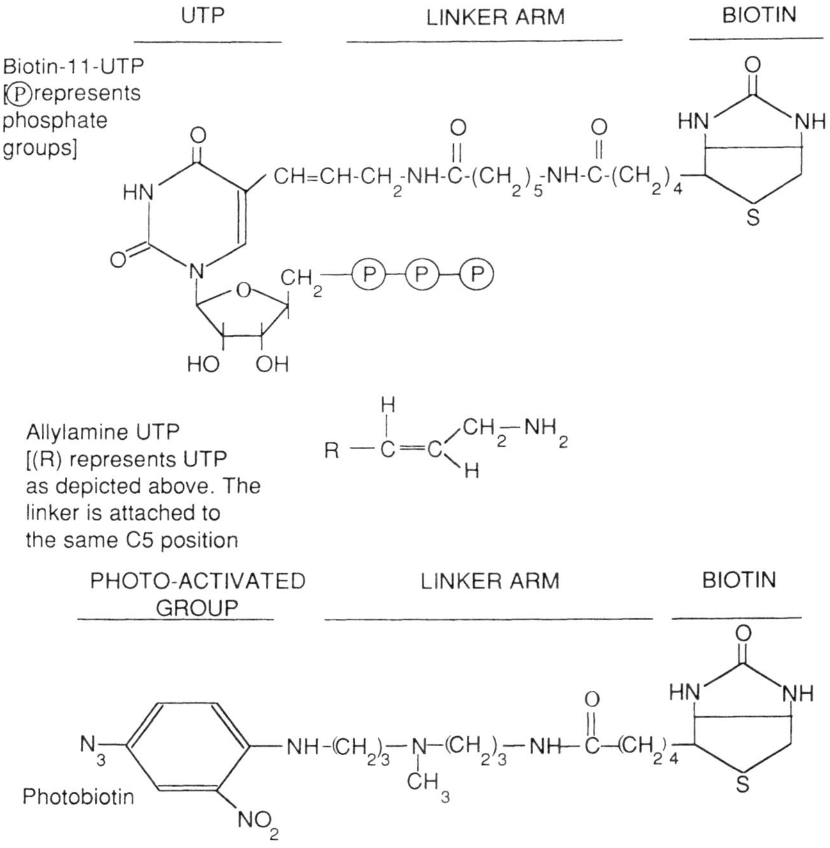

Figure 4 Schematic representation of biotin-11-dUTP and photobiotin.

Langer *et al.* (1981) synthesized biotin substituted ribonucleotides (bio-UTP) and deoxyribonucleotides (bio-dUTP) and showed that they could be incorporated into nucleic acids by a number of RNA polymerases and DNA polymerases respectively. These nucleotides are substituted at the C5 position of the pyrimidine base and are linked to a biotin molecule via a linker arm as shown in Figure 4. An alternative way of labelling nucleic acids with biotin is to incorporate allylamine-UTP into probes (Leuchrsen and Baum, 1987). The amino group present at the end of this molecule can then be reacted with ε-caproylamido-biotin-*N*-hydroxysuccinimide to produce a molecule similar to that depicted for bio-UTP. Finally RNA and DNA may be directly labelled using photobiotin, a photoreactable molecule which covalently links biotin to nucleic acids and proteins (Forster *et al.*, 1985).

It is important that biotinylated probes are not phenol extracted as the hydrophobic biotin moiety and the linker arm will cause the probe to enter the phenolic phase. Substitution of nucleotides with biotin has a slight effect on melting temperature (T_m). When 1 in 4 nucleotides are substituted with biotin a drop in T_m of 5°C is observed (Langer *et al.*, 1981). A slight reduction in T_m has also been demonstrated for short internally biotinylated oligonucleotides (Cook *et al.*, 1988). Small changes in T_m alone should not have a major effect on the rate of hybridization (see section 5) or final signal (provided stringency washes are done below the reduced T_m). Finally biotinylation will increase the apparent size of DNA (Cook *et al.*, 1988; Langer *et al.*, 1981) or RNA (personal observation) probes on gels.

The protein avidin or streptavidin forms a stable non-covalent link with biotin and acts as a bridge between the label and various possible detection systems (section 6). This provides an efficient way of labelling and detecting nucleic acid probes to a level of 1 pg of DNA/RNA on dot blots. Biotinylated probes have been used successfully to detect RNA and DNA *in situ* by a number of groups (Bhatt *et al.*, 1988; Guiteny *et al.*, 1988; Hopman *et al.*, 1986; Larsson *et al.*, 1988; Lawrence, Villnave and Singer, 1988; Mifflin *et al.*, 1987; Singer *et al.*, 1987; Syrjänen *et al.*, 1988; Zabel and Schafer, 1988).

While biotinylated probes have been used for *in situ* hybridization one of the main bars to their sensitivity is the presence of endogenous avidin binding activity in cells. This may be due to a non-specific interaction with components in the cell or may be due to endogenous biotin within the cell particularly in liver, kidney and lymphoid tissue. Biotin's normal role within the cell is as a co-factor for carboxylase enzymes and is found covalently bound to these enzymes (McCormick and Wright, 1971). For immunocytochemical (ICC) studies endogenous biotin has been blocked by pre-treating slides with streptavidin and then with free biotin (Wood and Warnke, 1981). The harsh conditions of hybridization and stringency washing required for *in situ* hybridization after such a pre-treatment may prevent this approach from working. Matsumoto (1985) describes a method for reducing endogenous avidin binding for use with ICC. This step is rather inconvenient requiring immersion in methanol for 12–24 hours. Lawrence, Vilnave and Singer (1988) report that the simple step of staining with avidin in 4 × SSC solution rather than phosphate buffered saline reduces non-specific avidin binding by ten fold.

4.2.2 Others

While biotin is the most commonly used non-radioactive label, there are several other labels in use by various groups. The availability of a number of labels opens up the possibility of simultaneous detection of multiple targets in the cell. In addition they offer a means of detecting hybridization in tissues rich in biotin.

4.2.3 Acetyl amino fluorene

In an approach described by Landegent *et al.* (1984) DNA or RNA for use as probes is modified by reaction with *N*-acetoxy-2-acetylamino-fluorene (AAF). This reacts with guanosine residues to produce a modified molecule which can be recognized using a specific antibody. AAF is a carcinogen and extra care is needed when handling and disposing of it.

4.2.4 Mercuration

When DNA or RNA is mixed with mercury ions there is a very rapid complexing between them. This is followed by a slower covalent linkage of mercury to the C5 position of the pyrimidine bases uracil and cytosine. Hopman *et al.* (1986) have labelled probes with mercury in this way. Sulphydryl groups will bind mercury very efficiently (association constant 10^{16} M^{-1}) and so probes can be detected using a molecule with a sulphydryl group and a strong antigenic determinant such as a trinitrophenyl group. This can then be detected using antibodies against the antigen. One disadvantage of this method is the use of the toxic substances mercury and potassium cyanide. Hopman *et al.* (1986) claim sensitivities equivalent to biotin. They have also gone on to demonstrate simultaneous detection of biotinylated and mercurated probe by *in situ* hybridization (Hopman *et al.*, 1986).

4.2.5 Digoxigenin, bromodeoxyuridine, sulphonation

The following three approaches involve incorporation of modified nucleotides into probes and subsequent detection via a primary antibody coupled to a detection system.

Digoxigenin is a steroid hapten which is linked to deoxyuridine triphosphate. This is incorporated into DNA probes using DNA polymerase and has been used for *in situ* hybridization to detect 1–2 copies per cell of HPV 16 DNA in SiHa cells (Heiles *et al.*, 1988).

Bromodeoxyuridine is an analogue of TTP which has been used *in vivo* to label cells which are in S phase. It has also been used by Niedobitek *et al.* (1988) to label viral probe DNA for *in situ* hybridization of lung tissue. This label is then detected by a specific antibody coupled via a multi-layered detection system to alkaline phosphatase. This group claims sensitivities equivalent to biotinylated probes by *in situ* hybridization and a detection level of 0.5 pg of probe on dot blots.

Sulphonation involves the modification of cytidine within DNA or RNA using bisulphite and *O*-methylhydroxylamine to generate a molecule modified with a sulpho group at the C5 position (Poverenny *et al.*, 1979). This is then detected by a specific antibody.

4.2.6 Direct cross-linking of probes to enzymes

Renz *et al.* (1984) have developed a means by which peroxidase and alkaline phosphatase may be linked directly to a DNA probe. The protein is fused with the polymer polyethylene-imine (PEI). The PEI enables the protein to bind electrostatically to any polyanion such as DNA. The resulting complex is then cross-linked using glutaraldehyde. Sensitivities of 1–5 pg on dot blots are achieved (Amersham, 1987). Jablonski *et al.* (1986) have crosslinked alkaline phosphatase to short modified oligonucleotides using disuccinimidyl suberate as a crosslinker and have achieved sensitivities equal to or better than ^{32}P. The use of this type of probe for *in situ* hybridization is currently under investigation.

5. HYBRIDIZATION

5.1 Introduction

Since the discovery of the double stranded helical nature of DNA much has been learned of the factors affecting this stable non-covalent interaction and the knowledge gained has been applied to a whole range of applications including *in situ* hybridization. The two strands of nucleic acid are held together by hydrogen bonding between bases on opposite strands and this is referred to as *base pairing*. Base pairing will only occur between cytosine (C) and guanine (G) or between adenine (A) and thymine (T) and so opposite strands must have a complementary sequence for base pairing to occur (Thymine is replaced by uracil in RNA molecules).

Denaturation describes the process by which a double stranded nucleic acid molecule separates into two single strands at elevated temperature. The temperature at which the strands are half dissociated is referred to as the *melting temperature* (T_m) and provides a measure of the stability of a given helix. Hydrogen bonding is greater between G:C base pairs than A:T base pairs and so the relative ratio of G:C to A:T pairs affects the hybrid stability and so its T_m. Other factors affecting T_m will be discussed later.

Hybridization is the process whereby complementary strands of DNA or RNA come together to form a stable double helix. The first stage of this, *nucleation*, involves an initial association between two short complementary regions. This is followed by a rapid association of adjacent sequences and has been called "*zippering*".

5.2 Factors affecting rate of hybridization

5.2.1 Probe concentration

Much of our knowledge of the factors affecting hybridization rates and specificities has been gained from solution hybridization and to a lesser extent from filter bound hybridization of DNA:DNA or DNA:RNA. The rate of hybridization of single stranded probes to target which is immobilized or in solution should follow pseudo first order kinetics if the probe concentration is greatly in excess of target. Thus the hybridization time will be inversely proportional to probe concentration (Meinkoth and Wahl, 1984; Birnstiel *et al.*, 1972). The situation appears to be more complex for *in situ* hybridization. Using single stranded RNA probes on glutaraldehyde fixed sea urchin embryos, Cox *et al.* (1984) found that the rate of hybridization in the first few hours was higher with increased probe concentration as expected. They also found however that the level of hybridization reached a plateau after 2–4 hours at all concentrations and before saturation of target sequence (even with a vast excess of probe). We have found a similar effect using paraformaldehyde fixed mouse fibroblasts (McCafferty *et al.*, 1989). The reason for this pattern is not clear.

When double stranded probes are used on immobilized targets, hybridization to target has to compete against the efficient re-annealing of probe in solution. This is a particular problem for *in situ* hybridization where diffusion of probe into the cell matrix has to occur before hybridization to mRNA. An inverse relationship between hybridization time and probe concentration has been found for *in situ* hybridization with double stranded DNA probes but the levels of hybridization achieved are 8 fold down on those reached with single stranded RNA probes (Cox *et al.*, 1984).

The situation is further complicated when double stranded DNA is the target. In this case denatured target may re-anneal and prevent hybridization of probe. The proximity of the two denatured strands in fixed tissue makes this increasingly likely. Singer *et al.* (1987) have found that hybridization reaches 1/3 of maximal levels in ten minutes and is complete within four hours. At this time re-annealing of the target DNA is probably complete. In order to combat the effects of self re-annealing of target it appears that the use of high probe concentration for a short time is preferable to using low probe concentrations for longer hybridization times.

5.2.2 Temperature

The rate at which DNA hybridization occurs is affected by temperature. In solution at 0°C the rate of hybridization is very low. As the temperature is raised the rate increases and a broad optimal range of hybridization temperatures spanning 10°C is found centred around a temperature which is 25°C below the T_m. As the hybridization temperature approaches the T_m, the rate drops off towards zero (Birnstiel *et al.*, 1972; Anderson and Young, 1985). A similar relationship for *in situ* hybridization hybrids has been reported (Cox *et al.*, 1984). This profile is found for perfectly matched hybrids and for mismatched hybrids. The latter however will have a lower T_m which is dependent on the degree of mismatch. Even at optimal temperatures the hybridization rate of mismatched molecules is less than for perfectly matched hybrids. For this reason a degree of specificity can be achieved by choosing the correct hybridization temperature. In practice most workers achieve selectivity by using post hybridization washes at temperatures approaching the T_m of the desired hybrid. It should be noted that Cox *et al.* (1984), using RNA probes, report a reduction of 5°C in the T_m of hybrids formed *in situ* when compared with hybrids in solution. Using short oligonucleotide probes, Lewis *et al.* (1985) have found no change in T_m. The fall in T_m for RNA probes and the apparent discrepancy between the two groups could be explained if shorter hybrids are formed *in situ* because of the fixed, protein bound nature of the target mRNA.

5.2.3 Salt concentration

Hybridization is most often carried out at salt concentrations in excess of 1.2 M NaCl. At concentrations below this the hybridization rate drops. This is particularly pronounced at salt concentrations below 0.1 M NaCl where a two fold drop in concentration can lead to a 5–10 fold drop in hybridization rate (Anderson and Young, 1985). While a high salt concentration increases the hybridization rate it also stabilizes mismatched hybrids. For this reason washes of hybrids are carried out in low salt concentrations (e.g. 30 mM) when a high degree of specificity is required.

5.2.4 Probe length

Studies of DNA re-annealing in solution reveal that probe length has a significant effect on melting temperature for probes up to several hundred base pairs. The effect becomes less pronounced as the probe size increases.

Increased probe sizes have also been shown to enhance the hybridization rate. In particular the rate of re-association is proportional to the square root of the probe size (Birnstiel *et al.*, 1972). Thus the longer length of probe provides more opportunity for the rate limiting nucleation event to occur. The situation becomes more complex for filter bound nucleic acids. When the concentration of probe is high in comparison to the target bound to the filter, the rate of hybridization is independent of probe length (Birnstiel *et al.*, 1972; Anderson and Young, 1985). When the situation is reversed and the amount of filter bound target is relatively high, hybridization becomes diffusion dependent and the rate is inversely proportional to the probe size i.e. smaller probes diffuse more easily.

For *in situ* hybridization it is very likely that probe diffusion is an important factor in determining the overall hybridization rate. The kinetics of hybridization have not been studied to the same extent as for solution or filter-bound hybridization. Nonetheless many labs reduce the size of probe molecules prior to hybridization (Cox *et al.*, 1984, Lynch *et al.*, 1987). In our own experience (McCafferty *et al.*, 1989) hydrolysis of RNA probes to a size of 100–250 bases gives best results (for paraformaldehyde fixed cells without pre-treatments, hybridized for 1 hour).

5.2.5 Formamide

Formamide and other denaturing agents such as urea destabilize double stranded nucleic acids. Thus the apparent T_m of hybrids will be lower in the presence of these agents. The T_m for DNA:DNA hybrids decrease linearly with increasing concentrations of formamide. Formamide reduces the T_m of DNA:RNA hybrids in a non-linear fashion and with less effect (Anderson and Young, 1985; Birnstiel *et al.*, 1972; Casey and Davidson, 1977; Cox *et al.*, 1984). The reduction in T_m means that hybridization and stringency washing can be done at lower temperatures. The practical implication of this is that strand scission and degradation of nucleic acid is reduced. In addition loss of target from filters or tissues will be reduced. Finally the lower temperature for *in situ* hybridization means that there is less chance of tissue morphology being compromized or of sections being lost from slides.

5.2.6 Dextran sulphate

The rate of hybridization may be increased by the inclusion of large anionic polymers such as dextran sulphate. These may work by excluding the probe from the volume occupied by the polymer and effectively increasing the probe concentration. It has also been suggested that network formation (discussed in Section 3) is enhanced when dextran sulphate is present.

6. DETECTION

6.1 Radioactive probes

Radioactive probes used for *in situ* hybridization are generally detected using autoradiographic emulsions. (As mentioned earlier, signal from ^{32}P labelled probes may be previewed by X-ray film before dipping in emulsion). The emulsions are essentially a dispersion of silver halide crystals in a gelatin matrix. These crystals are generally of uniform size and sensitivity. In response to light or ionizing radiation a cluster of silver atoms known as a latent image centre is formed within individual crystals. Upon development all crystals containing a latent image centre will be reduced to metallic silver to produce a visible silver grain.

Emulsions suitable for use include Amersham's LM-l, Ilford's K5 and Kodak's NTB2. Manufacturers recommendations should be followed for development, stopping and fixation.

Emulsions should be stored refrigerated (not frozen) away from ionizing radiation and chemicals which will react with the silver halide crystals (e.g. ammonia, sodium sulphide, formaldehyde and some organic solvents).

Before use the emulsion is melted and diluted if necessary. It is important to have an even thickness of emulsion on the slides and this is achieved by dipping the slides and withdrawing them from the dipping chamber at a uniform speed. The slides are allowed to set and dry before storing at 4°C with silica gel. Rapid drying of slides or sudden changes in temperature at any stage may cause stress throughout the emulsion which will in turn lead to higher background levels of silver grains. A more detailed coverage of all aspects of autoradiography is given by A W Rogers (1987).

Development of emulsions usually occurs in alkaline conditions and the reaction may be stopped in water or a dilute solution of acid. Fixation involves the removal of residual ionic silver halide. Sodium or ammonium thiosulphate is generally used for fixation to form a water soluble complex with silver halide which can be washed away.

While silver grains can be visualized using normal light field microscopy several labs use dark field microscopy for improved definition. A comparison of light field and dark field microscopy is given in Figure 3.

6.2 Non-radioactive probes

6.2.1 Detection of biotin

Biotin is the most commonly used non-isotopic label for *in situ* hybridization, and numerous different protocols are cited for its detection. Generally, however, all methods depend on the interaction of biotin with either the proteins avidin or streptavidin, or with a suitable antibody.

Avidin is a basic glycoprotein found in egg white, which has a very high affinity for biotin (Wichek and Bayer, 1988). One molecule can bind four biotin molecules and so link biotinylated probes to appropriate reporter molecules such as enzymes and fluorochromes. Due to the presence of oligosaccharide residues and its high isoelectric point, avidin tends to stick to tissue components, thereby increasing background. As a result, avidin is commonly replaced by streptavidin, a protein extracted from *Streptomyces avidinii*. This molecule also has four high affinity biotin binding sites, but lacks the "stickiness" of avidin (Cogg *et al.*, 1986). As avidin and streptavidin are used in exactly the same way, they shall hereafter be referred to as "avidin".

Only one binding site on each avidin molecule can interact with biotin on the hybridized probe. This leaves three sites available to bind to biotinylated enzymes (usually peroxidase or alkaline phosphatase). The enzymes can then generate a colorimetric signal through the precipitation of insoluble enzyme products. In one system alkaline phosphatase utilizes the substrates nitro blue tetrazolium (NBT) and 5-bromo-4-chloro-3-indolyl phosphate (BCIP) and produces a dark blue precipitate. Peroxidase utilizes diaminobenzidine (DAB) to produce a brown precipitate. Gold-silver amplification of the DAB signal may also be employed. This works by depositing silver on the enzyme reaction product so enabling better visualization of the signal (Shivers, *et al.*, 1986). Signals produced enzymatically can be monitored under the light microscope, and the reaction stopped when sufficient signal has accumulated.

The biotinylated enzymes may be added following the application of avidin (bridge system) or the avidin and enzyme may be added together as a preformed complex (Unger *et al.*, 1986). Alternatively, the complex may be chemically cross-linked by glutaraldehyde, to produce an avidin-enzyme conjugate. The use of complexes and conjugates reduces the number of steps in the detection protocol, but we have found that it also results in lower signal. This is probably due to poorer penetration into tissues. We therefore prefer to use the bridge system.

As an alternative to enzymes, fluorochromes such as fluorescein and rhodamine may be employed to identify the sites of hybridization. They are usually conjugated directly to avidin and function by producing a distinctive fluorescence when irradiated with light of the appropriate wavelength. For example, fluorescein produces a green fluorescence, and rhodamine fluoresces red. Fluorochromes provide better resolution than enzymes and have recently (Lawrence *et al.*, 1988) been shown to give equivalent sensitivities.

Biotin may be detected using anti-biotin antibodies as an alternative to avidin. In addition, such antibodies may be used in conjunction with avidin in

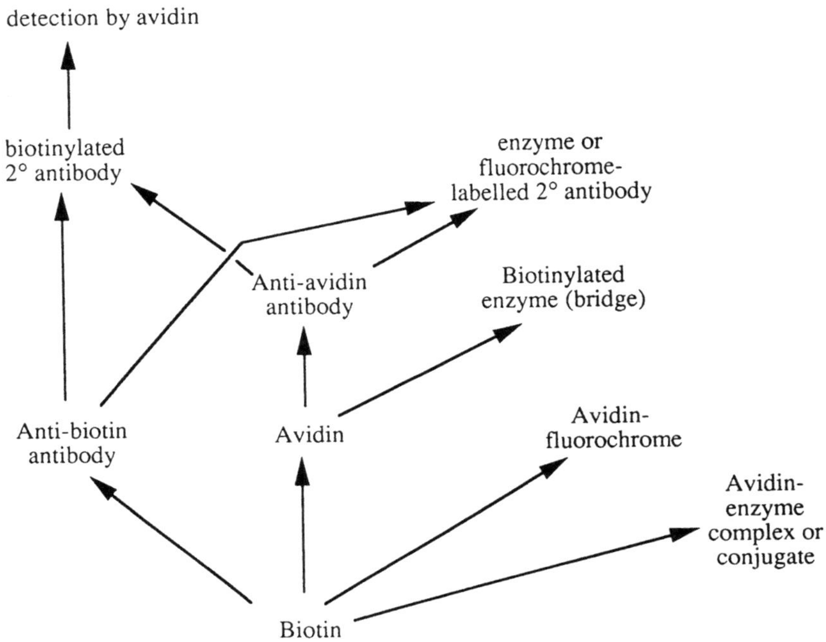

Figure 5 Schematic diagram indicating the various approaches available for the detection of a biotinylated probe.

order to amplify the signal. The indirect antibody method (van Noorden, 1986) is most frequently used. Here an anti-biotin antibody is bound to the biotinylated probe, and is then visualized with a secondary antibody, raised to the first, which is labelled with an enzyme or fluorochrome. As an alternative, the second antibody may be biotinylated, enabling detection with avidin via one of the methods described above. Since each secondary antibody may be biotinylated at a number of sites, amplification of the final signal can be achieved. Anti-avidin antibodies are also available (McDougall *et al.*, 1988), which can again be visualized with a secondary labelled antibody. The methods available for biotin detection are summarized in Figure 5.

6.2.2 Detection of other non-isotopic labels

Most other non-isotopic methods involve detection of modified nucleotides incorporated into probes. These are detected using antibodies generated against the labels. There are, for example, antibodies to guanosine-AAF (Landegent *et al.*, 1984) and bromodeoxyuridine (section 4.2.5). Mercury labelled probes are detected by utilizing an intermediate compound containing a mercury binding site and a strong antigenic determinant (Hopman *et al.*, 1986).

Once bound, the primary antibodies may be visualized by the methods already described in section 6.2.1.

7. CONTROLS FOR *IN SITU* HYBRIDIZATION

During the *in situ* hybridization procedure a variety of factors may contribute towards a spurious positive or negative result. To account for this, a number of controls should be included. The most commonly used controls are discussed below and the reader should select those which are most appropriate to his or her work.

7.1 Probe specificity

Prior to commencing an *in situ* hybridization experiment with an untried probe or tissue, it is advisable to check the probe's specificity on Northern or Southern blots. This will determine whether the probe hybridizes to a sequence of the expected size, without cross-reacting with any other nucleic acid species (Hoefler, 1987). If spurious bands do occur, then increasing the stringency of the post-hybridization washes may effect their removal. Failing this, a probe derived from a different region of the same gene, or subcloned from the original sequence may circumvent the problem. Signal distribution may in fact be verified by use of a second probe homologous to a different region of the same target sequence (Hoefler, 1987).

7.2 Non-specific probe binding

Probably the most common cause of false *in situ* hybridization signals is non-specific binding of the probe. There are both probe and tissue controls to assess this problem. If RNA probes generated from a dual promoter vector are used, then the sense strand (section 3.3) provides an ideal control for determining non-specific probe binding, having an identical specific activity, GC content and length as its antisense counterpart (Figure 2). If sense transcripts cannot be generated or double stranded DNA is used, then a probe lacking homology with sequences in the target tissue can be used. This should be checked using Northern or Southern blots. Vector sequences, if present, may result in non-specific signal. The extent of this problem can be determined by including labelled vector DNA as a control probe. If possible, vector sequences should be removed.

In the case of mRNA detection, a tissue that does not express the target sequence can be used as an ideal control for non-specific probe binding (Figure 2). Northern blotting will determine the presence or absence of target expression. In order to show that the probe is hybridizing to DNA or RNA, the target tissue may be treated with DNase and RNase respectively prior to hybridization. A potential drawback with this control is that signal reduction may be due to probe degradation caused by residual nucleases persisting until hybridization. Where DNA is the target, omission of the denaturation step will provide a control for non-specific binding.

7.3 Reproducibility

Controls are required to demonstrate the reproducibility of *in situ* hybridization experiments, and confirm that negative results are real and not due to faults in the procedure. This may be achieved by inclusion of a positive tissue, known to express the target sequence. A useful internal control is built into tissues with expressing and non-expressing regions (Lum, 1986). In addition, a probe may be used that is homologous to a sequence present in the target tissue. Actin is an excellent example of such a probe, being ubiquitously expressed in high copy numbers.

7.4 Autoradiographic detection

Spurious *in situ* hybridization results are frequently a consequence of the detection system used. Backgrounds may be produced in autoradiographic emulsions by the development process, pressure, light, environmental radiation and chemography (a reaction between tissue components and the emulsion: Rogers, 1979). For each experimental exposure, three control slides should be dipped and developed in order to account for these factors. One slide should be blank; the other two should carry tissue sections or cells. One of these should be taken through the *in situ* hybridization procedure without the addition of probe (no probe control, see Figure 2) and the other should be dipped without hybridization (no hybridization control).

7.5 Non-radioactive detection

For non-radioactive detection, a slide should be taken through the full hybridization and detection procedure, but without the addition of probe. This will indicate the level of background produced by the detection system itself. A further series of no probe controls in which various components of the detection system are omitted will identify the precise source of such background. In routine experiments, where it has been demonstrated that there is

Table 2 ***In situ* hybridization controls**

Factor controlled	*Procedure*
Probe specificity:	Northern or Southern Blot
Non-specific probe binding:	Sense RNA probe Non-homologous probe Negative tissue Nuclease digestion of target No DNA target denaturation
Reproducibility:	Positive control tissue Positive control probe
Target distribution:	Combined *in situ* hybridization and immunocytochemistry Second homologous probe
Autoradiographic detection:	Dip: No probe slide Unprocessed tissue section Blank slide
Non-radioactive detection:	No probe Omission of various detection reagents

no detection system background or its cause is known, this second set of controls is unnecessary.

Table 2 summarizes controls commonly used for *in situ* hybridization experiments, and their function.

ACKNOWLEDGEMENTS

We would like to thank Christiane Whitehead for typing this manuscript and colleagues at Amersham International for helpful comments and discussion.

REFERENCES

Allen JM, Sasek CA Martin JB, and Heinrich G (1987) Use of complementary ^{125}I labelled RNA for single cell resolution by *in situ* hybridization. Biotechniques **5**: 774–777.

Amersham International plc. ECL gene detection system. Technical bulletin S/313/89/1.

Amersham International plc. Nucleic acid labelling. Technical Bulletin. S197/87.

Anderson ML and Young BD (1985) Quantitative filter hybridization. In: *Nucleic Acid Hybridization*. (Hames BD ed.), pp. 73–110. IRL Press.

Bandtlow CE, Heumann R, Schwab ME and Thoenen H (1987) Cellular localization of nerve growth factor synthesis by *in situ* hybridisation. EMBO J. **6**: 891–899.

Berger CN (1986) *In situ* hybridization of immunoglobulin specific RNA in single cells of the B lymphocyte lineage with radiolabelled DNA probe. EMBO J. **5**: 85–93.

Bhatt B, Burns J, Flamery D and McGee J (1988) Direct visualisation of single copy genes on banded metaphase chromosomes by non-isotopic *in situ* hybridization. Nuc. Acids Res. **16**: 3951–3961.

Birnstiel ML, Sells BH and Purdom IF (1972) Kinetic complexity of RNA molecules. J. Mol. Biol. **63**: 21–39.

Bloch B, Popovici T, LeGuellec D, Normand E, Chouham S, Guitteny AF and Bohlen P (1986) *In situ* hybridization histochemistry for the analysis of gene expression in the endocrine and central nervous system tissues. J. Neurosci. Res. **16**: 183–200.

Brahic M and Haase AT (1978) Detection of viral sequences of low reiteration frequency by *in situ* hybridization. Proc. Natl. Acid. Sci. USA **75**: 6125–6129.

Brahic M, Haase AT and Cash E (1984) Simultaneous *in situ* detection of viral RNA and antigens. Proc. Natl. Acad. Sci. USA **81**: 5445–5448.

Brigati DJ, Myerson D, Leary JJ, Spalholz B, Travis SZ, Fong CKY, Hsiung AD and Ward DC (1983) Detection of viral genomes in cultured cells and paraffin-embedded tissue sections using biotin-labelled hybridization probes. Virology **126**: 32–50.

Burns J, Graham AK and McGee JO'D (1988) Non-isotopic detection of *in situ* nucleic acid in cervix: an updated protocol. J. Clin. Pathol. **41**: 897–899.

Busch MP, Rajagopalan MS, Gantz DM, Steimer KS and Vyas GN (1987) *In situ* hybridization and immunocytochemistry for improved assessment of human immunodeficiency virus cultures. Am. J. Clin. Pathol. 88: 673–680.

Butler ET and Chamberlin MJ (1982) Bacteriophage SP6 specific RNA polymerase. J. Biol. Chem. **257**: 5772–5778.

Casey J and Davidson N (1977) Rates of formation and thermal stabilities of RNA:DNA and DNA:DNA duplexes at high concentrations of formamide. Nuc. Acids Res. **4**: 1539–1552.

Chamberlin M, Kingston R, Gilman M, Wiggs J and deVera A (1983) Meth. Enzymol. **101**: 540.

Chien A, Edgar DB and Trela JM (1976) Deoxyribonucleic acid polymerase from the extreme thermophile *Thermus aquaticus*. J. Bacteriol. **127**: 1550–1557.

Cogg G, Del 'Orto P and Viale G (1986) Avidin-Biotin Methods. In: *Immunocytochemistry-Modern Methods and Applications*. (Polak JM, van Noorden S eds.), pp. 54–70, Wright, Bristol.

Cook AF, Vuocolo E and Brakel CL (1988) Synthesis and hybridization of a series of biotinylated oligonucleotides. Nuc. Acids Res. **16**: 4077–4095.

Costantini RM, Escot C, Theillet C, Gentile A, Merlo G, Liderean R and Callahan R (1988) *In situ* c-myc expression and genomic status of the c-myc locus in infiltrating ductal carcinoma of the breast. Cancer Res. **48**: 199–205.

Cox KH, DeLeon DV, Angerer LM and Angerer RC (1984) Detection of mRNA in sea urchin embryos by *in situ* hybridization using a symmetric RNA probes. Dev. Biol. **101**: 485–502.

Cumming RDF and Fallon RA (1988) Subcellular localization of biological molecules. In: *Radiosotopes in Biology*. (Slater R ed.) IRL Press.

Feinberg AP and Vogelstein B (1983) A technique for radiolabelling DNA restriction endonuclease fragments to high specific activity. Anal. Biochem. **132**: 6–13.

Feinberg AP and Vogelstein B (1984) Addendum: A technique for radiolabelling DNA restriction endonuclease fragments to high specific activity. Anal. Biochem. **137**: 266–7.

Forster AC, McInnes JL, Skingle DC and Symons RH (1985) Non-radioactive hybridization probes prepared by the chemical labelling of DNA and RNA with a novel reagent, photobiotin. Nuc. Acids Res. **13**: 745–761.

Gall JG and Pardue ML (1969) Formation of RNA:DNA hybrid molecules in cytological preparation. Proc. Natl. Acad. Sci. USA **63**: 378–383.

Godard CM (1983) Improved method for detection of cellular transcripts by *in situ* hybridization. Histochemistry **77**: 123–131.

Goedert M (1986) Single stranded DNA probes using an M13 template. In: *In Situ Hybridization in Brain*. (Uhl G ed), pp. 236–237, Plenum Press, New York.

Gomes SA, Nascimento JP, Siqueira MM, Krawsczuk MM, Pereira HG and Russell WC (1985) *In situ* hybridization with biotinylated DNA probes: a rapid diagnostic test for adenovirus upper respiratory infections. J. Virol. Meth. **12**: 105–110.

Green MR, Maniatis T and Melton DA (1983) Human β-globin pre-mRNA synthesised *in vitro* is accurately spliced in Xenopus oocyte nuclei. Cell **32**: 681–689.

Guitteny BF, Fouque B, Mongin C, Teoule R and Boch B (1988) Histological detection of mRNAs with biotinylated synthetic oligonucleotide probes. J. Histochem. Cytochem. **36**: 563–571.

Haase AT, Brahic M, Stowring L and Blum H (1984) Detection of viral nucleic acids by *in situ* hybridization. In: *Methods in Virology*. (Maramorosch K, Koprowski H eds.), VII pp. 189–226, New York Academic Press.

Hafen E, Levine M, Garber RL and Gehring WJ (1983) An improved *in situ* hybridization method for the detection of cellular RNA's in Drosophila tissue sections and its application for localising transcripts of the homeotic Antennapedia gene complex. EMBO J. **2**: 617–623.

Hayashi S, Gillam IS, Delaney AD and Tener GM (1978) Acetylation of chromosome squashes of Drosophila melanogaster decreases the background in autoradiographs from hybridization with ^{125}I labelled RNA. J. Histochem. Cytochem. **26**: 677–679.

Heiles MBJ, Genersch E, Kessler C, Neumann R and Eggers HJ (1988) *In situ* hybridization with digoxigenin-labelled DNA of human papillomaviruses (HPV 16/18) in HeLa and SiHa cells. Biotechniques **6**: 978–981.

Hoefler H, Childers H, Montminy MR, Lechan RM, Goodman RH and Wolfe HJ (1986) *In situ* hybridization methods for the detection of somatostatin mRNA in tissue sections using anti-sense RNA probes. Histochem. J. **18**: 597–604.

Hoefler H (1987) What's new in *in situ* hybridization. Path. Res. Pract. **182**: 421–430.

Hopman AHN, Wiegant J and van Duijn P (1986) A new hybridocytochemical method based on mercurated nucleic acid probes and sulphydryl-hapten ligands. Histochemistry **84**: 169–185.

Hopman AHN, Wiegant J, Raap AK, Landegent JE, van der Ploeg M and van Duijn P (1986) Bi-colour detection of two target DNA's by non-radioactive *in situ* hybridization. Histochemistry **85**: 1–4.

Hu N and Messing J (1982) The making of strand specific M13 probes. Gene **17**: 271–277.

Jablonski E, Moomaw EW, Tullis RH and Ruth JL (1986) Preparation of oligonucleotide-alkaline phosphatase conjugates and their use as hybridization probes. Nuc. Acids Res. **14**: 6115–6128.

Johnson SA, Morgan DG and Finch CE (1986) Extensive postmortem stability of RNA from rat and human brain. J. Neurosci. Res. **16**: 267–280.

Landegent JE, Jansen in de Wal N, Baan RA, Hoeijmakers JHJ and van der Ploeg M (1984) 2-Acetylaminofluorene-modified probes for the indirect hybridocytochemical detection of specific nucleic acid sequences. Exp. Cell Res. **153**: 61–72.

Langer PR, Waldrop AA and Ward DC (1981) Enzymatic synthesis of biotin labelled polynucleotides: Novel nucleic acid affinity probes. Proc. Natl. Acad. Sci. USA **78**: 6633–6637.

Larsson LJ, Christensen T and Dalboge H (1988) Detection of POMC mRNA by *in situ* hybridisation using a biotinylated oligodeoxynucleotide probe and avidin-alkaline phosphatase histochemistry. Histochemistry **89**: 109–116.

Lawrence JB and Singer RH (1985) Quantitative analysis of *in situ* hybridization methods for detection of actin gene expression. Nuc. Acids Res. **13**: 1777–1799.

Lawrence JB, Villnave CA and Singer RH (1988) Sensitive high resolution chromatin and chromosome mapping *in situ*: presence and orientation of two closely integrated copies of EBV in a lymphoma line. Cell **52**: 51–61.

Lewis ME, Sherman TG and Watson SJ (date) *In situ* hybridization histochemistry with synthetic oligonucleotides: strategies and methods. Peptides **6**: Suppl. 2: 75–87.

Lewis ME, Arentzen R, and Baldino F (1986) Rapid, high resolution *in situ* hybridization histochemistry with radioiodinated synthetic oligonucleotide. J. Neurosci. Res. **16**: 117–124.

Luechrsen KR and Baum MP (1987) *In vitro* synthesis of biotinylated RNA probes from A-T rich templates: Problems and solutions. Biotechniques **5**: 660–670.

Lum JB (1986) Visualization of mRNA transcription of specific genes in human cells and tissues using *in situ* hybridization. Biotechniques **4**: 32–38.

Lynch KR, Hawelu-Johnson CL and Guyenet PG (1987) Localization of brain angiotensinogen mRNA by hybridization histochemistry. Mol. Brain Res. **2**: 149–158.

Maniatis T, Fritsch EF and Sambrook J (1982) *Molecular Cloning*, New York, Cold Spring Harbor Laboratory.

Matsumoto Y (1985) Simultaneous inhibition of endogenous avidin-binding activity and peroxidase applicable for the avidin biotin system using monoclonal antibodies. Histochemistry **83**: 325–330.

McCafferty J, Cresswell L, Alldus C, Terenghi G and Fallon R (1989) A shortened protocol for *in situ* hybridization to mRNA using radiolabelled RNA probes. Technique **1**: 171–182 and erratum (1990) **2**: 112.

McCormick DB and Wright LD (1971) The metabolism of biotin and analogues. Comprehensive Biochem. **21**: 81–109.

McDougall JK, Myerson D and Beckmann AM (1988) Detection of viral DNA and RNA by *in situ* hybridization. J. Histochem. Cytochem. **34**: 33–38.

Meinkoth J and Wahl G (1984) Hybridization of nucleic acids immobilised on solid supports. Anal. Biochem. **138**: 267–284.

Melton D, Kneg P, Rebagliati M, Maniatis T, Zinn K and Green MR (1984) Efficient *in vitro* synthesis of biologically active RNA and RNA hybridization probes from plasmids containing a bacteriophage SP6 promoter. Nuc. Acids Res. **12**: 7035–7056.

Mifflin TE, Bowden J, Lovell MA, Bruns DE, Hayden FG, Groschel DHM and Savory J

(1987) Comparison of radioactive and biotinylated probes for detection of cytomegalovirus DNA. Clin. Biochem. **20**: 231–235.

Moench TR, Gendelman HE, Clements JE, Narayan O and Griffin DE (1985) Efficiency of *in situ* hybridization as a function of probe size and fixation technique. J. Virol. Methods **11**: 119–130.

Morley DJ and Hodes ME (1987) *in situ* localization of amylase mRNA and protein. An investigation of amylase gene activity in normal human parotid gland. J. Histochem. Cytochem. **35**: 9–14.

Niedobitek G, Finn T, Herbst H, Bornoft G, Gerdes J and Stein H (1988) Detection of viral DNA by *in situ* hybridization using bromodeoxyuridine-labelled DNA probes. Am. J. Pathol. **131**: 1–4.

van Noorden S (1986) Tissue preparation and immunostaining techniques for light microscopy In: *Immunocytochemistry – Modern Methods and Applications.* (Polak JM, van Noorden S eds.), 2nd edition, Wright, Bristol.

Poverenny AM, Podgorodnichenko VK, Bryksina LE, Monastyrskayo GS, and Sverdlov ED (1979) Immunochemical identification of the product of cytosine modification with bisulphite and O-methylhydroxylamine mixture. Mol. Immunol. **16**: 313–316.

Renz M and Kurz C (1984) A colorimetric method for DNA hybridization. Nuc. Acids Res. **12**: 3435–3444.

Rigby PW, Dickman M, Rhodes C, and Berg P (1977) Labelling deoxyribonucleic acid to a high specific activity *in vitro* by nick translation with DNA polymerase. J. Mol. Biol. **113**: 237.

Rogers AW (1979) *Techniques of Autoradiography.* (Elsevier/North Holland Biomedical Press).

Shivers BD, Harlan RE, Pfaff DW and Schachter BS (1986) Combination of immunocytochemistry and *in situ* hybridization in the same tissue section of rat pituitary. J. Histochem. Cytochem. **34**: 39–43.

Singer RH and Ward DC (1982) Actin gene expression visualised in chicken muscle tissue culture by using *in situ* hybridization with a biotinylated nucleotide analogue. Proc. Natl. Acad. Sci. USA **79**: 7331–7335.

Singer RH, Lawrence JB and Rashtchian RN (1987) Toward a rapid and sensitive *in situ* hybridisation methodology using isotopic and non-isotopic probes. In: *In Situ Hybridization.* (Valentino KL ed.), Oxford University Press.

Singer RH, Lawrence JB and Villnave C (1986) Optimisation of *in situ* hybridization using isotopic and non-isotopic detection methods. Biotechniques **4**: 230–250.

Syrjänen S, Partanen P, Mantyjarvi R and Syrjänen K (1985) Sensitivity of *in situ* hybridization techniques using biotin and ^{35}S labelled human papillomavirus (HPV) DNA probes. J. Virol. Meth. **19**: 225–238.

Terenghi G, Polak JM Hamid Q, O'Brien E. Denny P, Legon S, Dixon J, Minth CD, Palay SL, Yasargil G and Chan-Palay V (1987) Localization of neuropeptide Y mRNA in neurons of human cerebral cortex by means of *in situ* hybridization with a complementary RNA probe. Proc. Natl. Acad. Sci. USA **84**: 7315–7318.

Tourner I, Bernuau D, Poliard A, Schoevaert D and Feldmann G, (1987) Detection of Albumin mRNAs in rat liver by *in situ* hybridization: paraffin embedding and comparison of fixation procedures. J. Histochem. Cytochem. **35**: 453–459.

Uhl GR, Evans J, Parta M, Walworth C, Hill K, Sasek C, Voigt M and Reppert S (1986) Vasopressin and somatostatin mRNA *in situ* hybridization. In: *In Situ Hybridization in Brain.* (Uhl GR ed.), Plenum Press, New York.

Unger ER, Budgeon LR, Myerson D and Brigati DJ (1986) Viral diagnosis by *in situ* hybridization. Am. J. Sur. Pathol. **10**: 1–8.
Venezky DL, Angerer LM and Angerer RC (1981) Accumulation of histone repeat transcripts in the sea urchin egg pronucleus. Cell **24**: 385–391.
Wetmur JG and Davidson N (1968) Kinetics of renaturation of DNA. J. Mol. Biol. **31**: 349–370.
Wilchek M and Bayer EA (1988) The avidin-biotin complex in bioanalytical applications. Anal. Biochem. **171**: 1-32.
Wood GS and Warnke R (1981) Suppression of endogenous avidin binding activity in tissues and its relevance to biotin avidin detection systems. J. Histochem. Cytochem. **29**: 1196–1204.
Zabel M and Schafer H (1988) Localization of calcitonin and calcitonin gene-related peptide mRNAs in rat parafollicular cells by hybridocytochemistry. J. Histochem. Cytochem. **36**: 543–546.
Zawatsky R, De Maeyer E and De Maeyer-Guignard J (1985) Identification of individual interferon producing cells by *in situ* hybridization. Proc. Natl. Acad. Sci. USA **82**: 1136–1140.

2
Use of comparative *in situ* hybridization and immunocytochemistry for the study of regulatory peptides

Giorgio Terenghi and Julia M. Polak
Histochemistry Department, Royal Postgraduate Medical School, Hammersmith Hospital, Du Cane Road, London W12 0NN

1. INTRODUCTION

In the past decade immunocytochemistry has revolutionized the approach to morphological studies of tissues, allowing the localization of specific peptides and other cellular markers to be demonstrated. However, the knowledge gained by identifying cellular antigens is limited, as simple localization cannot give any information on the dynamic changes of cellular metabolic processes. The recent introduction of *in situ* hybridization for mRNA or DNA using labelled complementary nucleic acid probes has made possible the understanding of gene expression at the cellular level in different physiological and pathological conditions. Compared to other morphological and molecular biological techniques, *in situ* hybridization offers precise anatomical localization and increased sensitivity, opening completely new horizons to the scope of morphological studies.

Regulatory peptides have been recognized by immunocytochemistry in many cell types in almost all tissues and organs (see Polak and Van Noorden, 1986; Polak, 1989 for review). Often the site of synthesis of any given peptide is specific to only one cell type within a heterogeneous tissue, and its mRNA may represent only a small fraction of the total cellular mRNA. The most common biochemical method for RNA identification, Northern blot analysis, requires extraction of the nucleic acid from homogenized tissue. If the target sequence is present only in a minority of the total cell population, there is a dilution effect which can result in a loss of sensitivity. Furthermore, this technique cannot provide an identification of the cell type containing the mRNA under study. Using *in situ* hybridization it is possible to localize anatomically any identified mRNA because of the specificity of complementary base pairing between nucleic acid probes and mRNA target. Also the sensitivity of *in situ* hybridization is considerable, with limit of detection down to single copy gene sequences (Bhatt *et al.*,1988; Lawrence *et al.*, 1988).

1.1 Choice of probe

When applying *in situ* hybridization, one of the important factors is the selection of a suitable probe. The alternatives are many, ranging from single- or double-stranded RNA or DNA probes, to whether the label should be radioactive or non-radioactive. It is beyond the scope of this chapter to examine the advantages and disadvantages of the different choices, which have already been reviewed elsewhere (Terenghi and Fallon, 1989), but it is clear that at present there is no universal system appropriate for all applications. The final choice is very much dependent on the system under investigation, and also on the sensitivity, speed and resolution required.

Although our initial studies on detection of peptide mRNA were carried out using single stranded DNA probes (Varndell *et al.*,1984), in our present studies we prefer to use RNA probes (Cox *et al.* , 1984) for several reasons.

RNA probes are single stranded and do not require denaturation prior to use, reannealing in solution does not occur and all the probe is available for hybridization. The melting temperature of RNA/RNA hybrids is higher than for DNA/RNA hybrids (Casey and Davidson, 1977; Wertmur, Ruyecham and Donthart, 1981; Cox *et al.*, 1984), hence higher temperatures can be used for hybridization and stringency washes to prevent or remove weak or non-specific probe binding. Since RNA probes are free from vector sequence, lower probe concentrations are required for saturation of target sequences, consequently a higher signal/noise ratio can be obtained, as background signal is also dependent on the concentration of the applied probe. RNA probes are easily synthesized using RNA polymerase (Melton *et al.*, 1984; Angerer *et al.*, 1985), with up to 80% incorporation of radiolabelled nucleotides and high specific activity. Biotinylated nucleotides can also be incorporated efficiently into RNA probes, with very good hybridization results (Giaid *et al.*,1989).

An interesting development has been the construction of RNA probe plasmids by inserting an oligonucleotide in a dual promoter vector (Wolf *et al.*, 1987). More recently, specific oligonucleotides have been produced which include a promoter sequence for RNA polymerase (Brysch *et al.*, 1988). These approaches combine the advantages of oligonucleotide probes with those of RNA probes, as it is possible to generate uniformly labelled cRNA probes from oligonucleotide templates. These probes have now been tried successfully for peptide *in situ* hybridization (Denny *et al.*, 1988).

1.2 Methodology

The main steps of the *in situ* hybridization procedure can be summarized as follows:

(i) tissue preparation, which includes fixation to ensure retention of target sequences and to maintain tissue morphology;

(ii) pre-hybridization treatments, to increase probe access to the target sequence and to reduce non-specific probe binding;

(iii) probe labelling with either radionucleotides or non-isotopic labels;

(iv) hybridization under optimal conditions;

(v) probe detection with either autoradiography or immunocytochemical methods, according to the type of label in use.

In our experience the approaches to radioactive and non-radioactive *in situ* hybridization are very similar, with variation of the main protocol restricted to differences in concentration of solutions or timing of different steps (cf. Terenghi *et al.*, 1987; Giaid *et al.*, 1989a). The choice of radiolabels is generally determined by a balance between speed and resolution, and in our experience ^{32}P and ^{35}S are best suited for studies of the neuroendocrine system. Liquid emulsion autoradiography is generally used, although film autoradiographs can be quantified densitometrically (Steel *et al.*, 1989: see below).

There is now a wide choice of methods for labelling nucleic acid probes non-isotopically (Mitchell *et al.*, 1986; Van der Ploeg *et al.*, 1986; Niedobitek *et al.*, 1988; Hopman *et al.*, 1986). Biotin is the most commonly used label, particularly for DNA probes, possibly because of the poor incorporation generally obtained by different investigators during biotinylation of RNA probes. However, in our experience RNA probes can be successfully labelled with biotin-UTP, obtaining an incorporation of up to 35–50% (Giaid *et al.*, 1989a). With an alternative approach using allylamine-UTP, which is then chemically reacted with biotin (Luehrsen, and Baum, 1987), label incorporation of up to 90% can be obtained.

Using either biotin-UTP or allylamine-UTP labelled probe a comparably strong hybridization signal with low background was achieved, for prolactin and growth hormone (GH) in pituitary sections. However, the choice of detection system was important in obtaining the best signal/noise ratio. A wide range of immunocytochemical methods can be used for the detection of hybridized biotinylated probes, but the best results were obtained when an avidin-biotin complex (ABC) peroxidase detection system was used in conjunction with glucose-oxidase-diaminobenzidine (DAB)–nickel solution (Figure 1; Giaid *et al.*, 1989a).

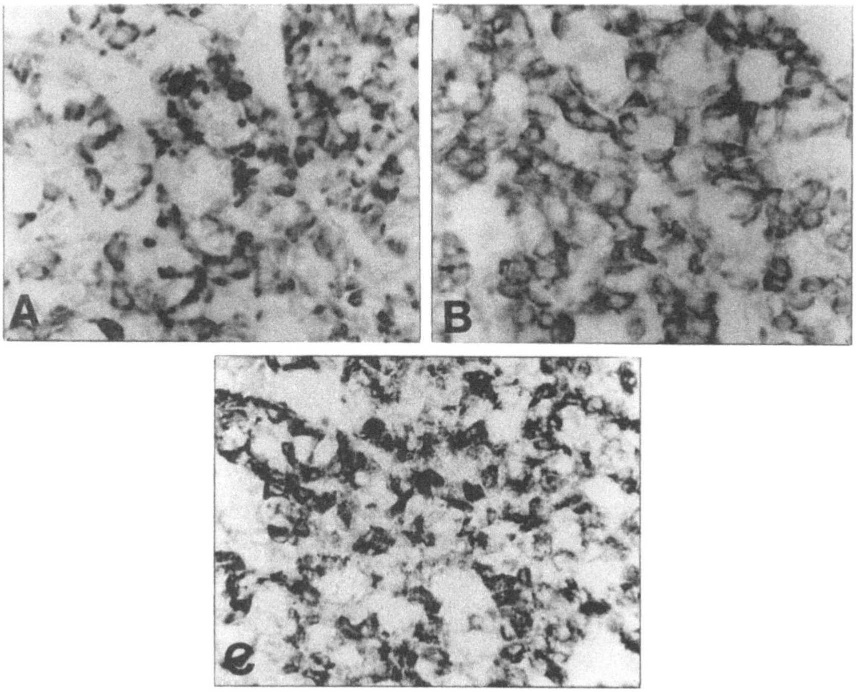

Figure 1 Sections of rat pituitary hybridized with biotinylated RNA probes encoding growth hormone (GH). Different signal/noise ratios are observed when using a variety of immunocytochemical detection systems. (**A**) ABC peroxidase method, development in glucose oxidase-DAB-nickel solution; (**B**) monoclonal anti-biotin antiserum followed by ABC peroxidase method, development as above; (**C**) method as (B), development in DAB enhanced with silver methenamine. (×200).

Although biotinylated probes have been criticized for lack of sensitivity and increased background noise compared to radiolabelled ones (Singer *et al.*, 1986; Gillam, 1987), more recent publications have shown that they can be used to detect even single copy genes (Bhatt *et al.*, 1988; Lawrence *et al.*, 1988), which indicates the enormous potential of non-radioactive *in situ* hybridization in future research and diagnostic techniques.

1.3 Immunocytochemistry and *in situ* hybridization

Immunocytochemistry is an established technique which can supply very useful information on cell identification and peptide content, although the latter may not correspond with the turnover rate or the metabolic state of the cell. The combined application of immunocytochemistry and *in situ* hybridization can provide more complete correlative data on peptide synthesis and storage in specific situations.

The two techniques can be carried out on separate preparations or serial sections according to established methods and the results compared (Terenghi *et al.*, 1987; Hamid *et al.*, 1987; Steel *et al.*, 1988a). Sequential application of immunocytochemistry and *in situ* hybridization on the same section has also been used by different groups to investigate mRNA synthesis and translated peptide/protein storage during viral infection (Brahic *et al.*, 1984) and in a variety of normal and pathological tissues (Wolfson *et al.*, 1985; Shivers *et al.*, 1986a; Hofler *et al.*, 1987; Chan-Palay *et al.*, 1988; Steel *et al.*, 1989). This approach offers considerable advantages as there is maximal anatomical accuracy in cell identification, which is particularly important when a homogeneous type of tissue is used that offers few recognizable landmarks; also there is no sampling error and the identification of the reaction products within the same cells is immediate (Colour plates 1A and 1B).

The sequential methods can be applied in two different ways: immunocytochemistry first, followed by hybridization, or hybridization first followed by immunostaining before the autoradiography step. In neither case do the techniques need major alteration, but there are a few practical considerations to take into account.

If immunocytochemistry is carried out first, all the reagents must contain an RNase inhibitor to avoid degradation of the nucleic acid target, and heparin has been used successfully for this purpose (Hofler *et al.*, 1987; Steel *et al.*, 1989). However, some loss of hybridization signal seems to occur because of partial mRNA degradation, as well as decreased immunostaining which may be due to impaired antigen-antibody binding caused by the RNase inhibitors (Shivers *et al.*, 1986a).

When immunocytochemistry is carried out after hybridization, there is the risk that the antigenic sites might be damaged by the reagents and/or temperatures required for hybridization (Brahic *et al.*, 1984; Shivers *et al.*, 1986a). In the past, omission of dextran sulphate from the hybridization buffer has been recommended, as it can bind to proteins and impair their antigenicity (Hofler *et al.*, 1987; Chan-Palay *et al.*, 1988). This omission can lower considerably the binding of probe, and thus decrease the hybridization signal (Hofler *et al.*, 1987). In recent studies, it has been demonstrated that the presence of dextran does not affect the subsequent immunostaining (Steel *et al.*, 1989), although it is to be remembered that the effect on different antigens might vary considerably.

2. APPLICATIONS

2.1 Anatomical localization of peptide synthesis

Regulatory peptides have been widely studied by immunocytochemistry. However, the demonstration of an intracellular peptide antigen does not confirm that the peptide is synthesized in that cell. *In situ* hybridization can supply such information, and sometimes demonstrate peptide gene expression in unexpected locations.

Atrial natriuretic peptide (ANP), which affects blood pressure, renal function and salt balance, had been identified by immunocytochemistry in both atria and ventricles of mammalian heart. Different studies have demonstrated the synthesis of peptides in atrial myocytes, but it was unclear whether the presence of ANP in ventricular cells was due to cell synthesis or uptake of circulating peptide. By using probes which specifically recognize the mRNA sequence of human and rat ANP, it has been possible to confirm ANP gene expression in both atrial and ventricular cells, in tissue sections and cultured myocytes (Hamid *et al.*, 1987; Figure 2).

ANP has also been found in other tissues, such as pituitary, kidney, adrenal gland, salivary gland, eye and brain. In lung, ANP has been detected by

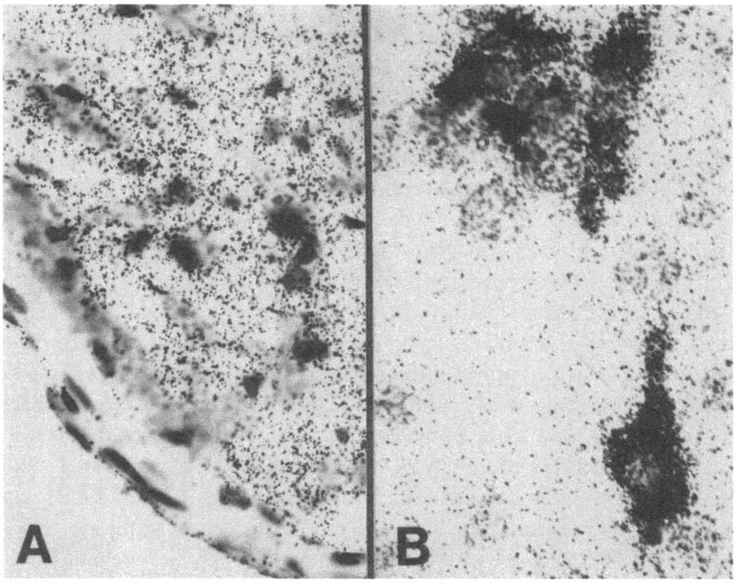

Figure 2 (**A**) Section of rat atrium (×200) and (**B**) cultured myocytes of rat atrium (×325) hybridized with ^{32}P-labelled ANP probe. Haematoxylin counterstain.

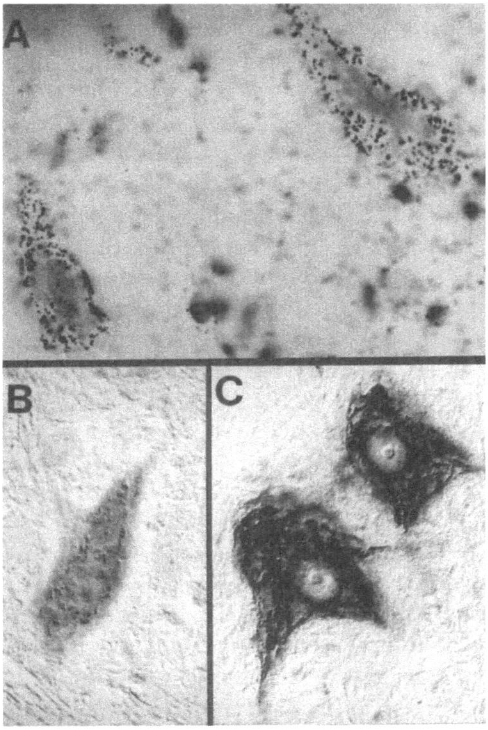

Figure 3 (**A**) Section through the lumbar ventral horn of human spinal cord hybridized with ^{35}S-labelled CGRP probe. Haematoxylin counterstain (×300). Motoneurons in sections of (**B**) human lumbar 28 ventral horn (×300) and (**C**) rat lumbar ventral horn (×340) immunostained with antisera to CGRP using the PAP method. Note the granular immunoproduct.

radioimmunoassay, and it was postulated that the peptide might be localized to parenchymal cells. For a long time it has been known that cardiac muscle extends to extrapulmonary veins and, particularly in rodents, also to intrapulmonary veins. By using a combination of *in situ* hybridization and immunocytochemistry it was possible to demonstrate that in rat ANP synthesis and storage occurs in muscle cells of both extra- and intra-pulmonary veins (Springall *et al.*, 1988). The functional significance of ANP in pulmonary veins is still unclear, but its pulmonary release, possibly caused by vein stretching, might have a local paracrine effect, as well as contributing to the levels of circulating peptide.

Using probes to different peptides, it has been possible to compare their transcription sites with the known distribution of the mature peptides. Calcitonin gene-related peptide (CGRP) is a peptide which has been localized by immunocytochemistry throughout the sensory nervous system. CGRP has been shown to be co-localized with the putative sensory neurotransmitter substance P (SP) (Gibson *et al.*, 1984; Terenghi *et al.*, 1985), and both peptides have been shown to mediate the injury response. Unexpectedly, CGRP was also immunostained in motoneurones in ventral horn of the spinal cord (Gibson *et al.*, 1984). CGRP-immunoreactive motoneurones often have a granular appearance, suggestive of CGRP-immunoreactive terminal synapsing on the motoneural soma, and raising doubts about the source of the peptide. With *in situ* hybridization it was possible to confirm that in both man and rat numerous sensory neurons of dorsal root ganglia show CGRP gene expression, and that mRNA is also present in motoneurons (Gibson *et al.*, 1988; Giaid *et al.*, 1989b; Figure 3). Probes for both α- and β-CGRP, which have a closely related sequence, were used to hybridize tissue preparations and Northern blots. A predominance of α-CGRP expression was clear, with greatest levels of expression in dorsal root ganglion, and somewhat less in spinal ventral horn. These findings have great importance in view of the recent reports which attribute motor related physiological actions to CGRP.

As another example, neuropeptide Y (NPY) gene transcription has been shown in neuronal cells of the human cerebral cortex (Terenghi *et al.*, 1987). The distribution of hybridized cells was consistent with that of NPY-immunoreactive neurons, although a small numerical difference was noted between cells identified with the two methods. This might be due to different metabolic states of the cells resulting in a pattern of gene transcription different from that of peptide storage. However, combined *in situ* hybridization and immunocytochemistry investigations have confirmed co-localization of transcribed mRNA and translated peptide (Chan-Palay *et al.*, 1988). The function of NPY-producing neurons is under investigation, as they show numerical and morphological changes in some neurological diseases (Chan-Palay *et al.*, 1988).

2.2 Localization of new peptides

Endothelin is a recently identified peptide, which has been isolated from endothelial cell cultures and shows a distinctive vasoconstrictor action (Inoue *et al.*, 1989). In addition, electrophysiological studies have shown a direct effect of endothelin on neuronal excitability, consistent with identified receptor binding sites in central and peripheral nervous systems. The possible localization of endothelin in the nervous system of man was investigated with both

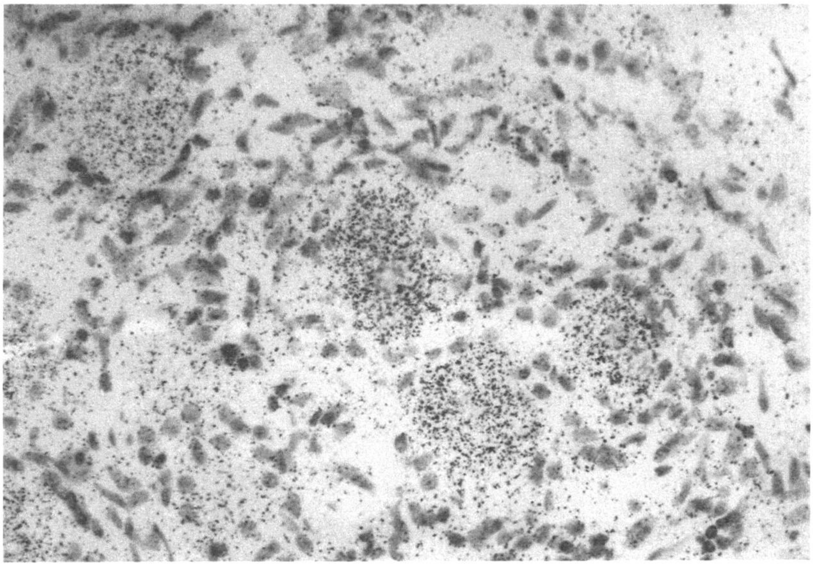

Figure 4 Section of human dorsal root ganglion hybridized with ^{35}S-labelled probe encoding endothelin. Haematoxylin counterstain (×300).

immunocytochemistry and *in situ* hybridization (Giaid *et al.*, 1989b). In dorsal root ganglia endothelin mRNA was localized to the majority of both large and small neuronal cell bodies (Figure 4). In the spinal cord, numerous positive cells were seen in both dorsal and ventral horns. Comparison of sections also showed that endothelin mRNA almost always co-exists with substance P and/or CGRP transcripts in subpopulations of neuronal cells, and consistent results were obtained with immunocytochemistry for the mature peptides. The co-localization also extends to motoneurons in the ventral horn of the spinal cord, where a subset of cells expresses both CGRP and endothelin.

These studies clearly demonstrated the previously undetected presence of endothelin in sensory and motor neurons of the human nervous system, strongly supporting a putative role of this newly discovered peptide as a neurotransmitter or neuromodulator. It is also interesting that endothelin is co-localized with substance P and CGRP at both synthesis and storage levels, as these two peptides are known to participate in the modulation of sensory and motor functions.

2.3 Functional changes in endocrine cells

The study of pituitary peptide-containing cells in different endocrine situations shows the advantages of combining *in situ* hybridization and immunocytochemistry in order to obtain more information on intracellular metabolism. The synthesis and secretion of prolactin was examined in animal models throughout pregnancy and lactation, and after ovariectomy as oestrogen are known to regulate prolactin turnover.

Immunostaining for prolactin showed no appreciable difference in the pituitaries of control, pregnant or lactating rats, although it is well documented that the level of circulating prolactin is very much altered during these physiological states. By using *in situ* hybridization it was possible to show that in control animals the hybridization signal for prolactin mRNA was present in a number of cells similar to that observed with immunocytochemistry. However, the frequency and signal intensity of the hybridized cells was decreased in pregnant animals as compared to controls (Steel *et al.*, 1988a). The suppression of prolactin synthesis during pregnancy, clearly evident by *in situ* hybridization, was in contrast with the unaltered immunoreactivity for the peptide, and suggested increased peptide storage and decreased secretion. Conversely, in lactating animals both prolactin synthesis and secretion were enhanced, resulting in increased hybridization signal but again unaltered immunoreactivity (Figure 5). In contrast, after ovariectomy both prolactin-immunoreactivity and prolactin hybridization signal were present in a comparable number of cells, although these were less numerous than in controls (Steel *et al.*, 1988a).

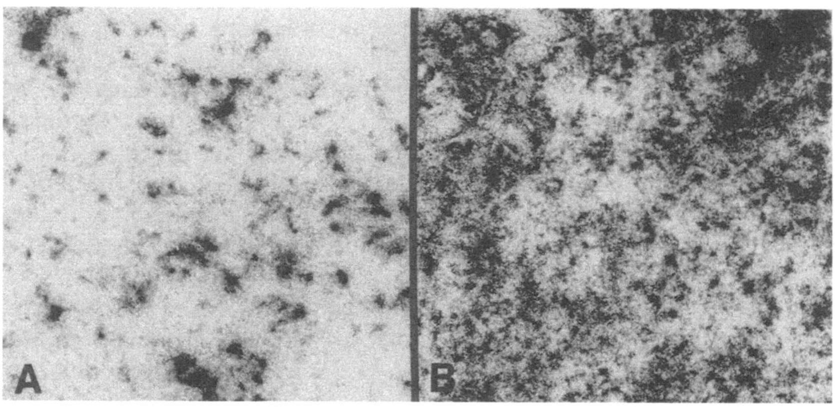

Figure 5 Pituitary sections of (**A**) control and (**B**) lactating rat hybridized with ^{32}P-labelled prolactin probe. Haematoxylin counterstain (×160).

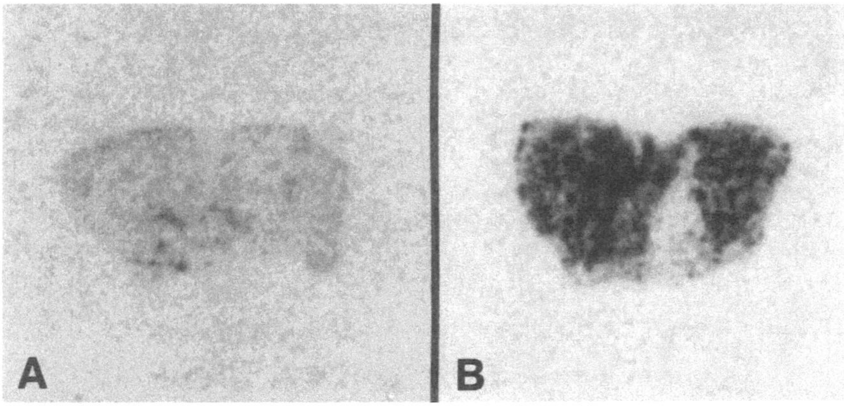

Figure 6 Film autoradiographs for densitometric quantification of pituitary sections of **A**) control and (**B**) thyroidectomized rat hybridized with ^{35}S-labelled β-TSH probe (×8).

In a separate study on peptide gene regulation, the expression of beta thyroid-stimulating hormone (β-TSH) was examined in rats under euthyroid and hypothyroid conditions, since TSH synthesis is regulated by thyroid hormones. In pituitaries of hypothyroid rats, the number and signal intensity of hybridized cells was greatly increased compared to that in euthyroid control rats (Steel *et al.*, 1989; Colour plates 1A and1B; Figure 6). This increased gene expression is clearly consistent with the absence of negative feedback control exerted by thyroid hormones on TSH synthesis. Similarly, in adrenalectomized rats there is increased pro-opiomelanocortin (POMC) gene expression which is correlated to the length of time after operation (Steel *et al.*, 1988b).

2.4 Quantification

When comparing changes of gene expression during physiological or pathological events, it is important to express the variation objectively in measurable quantities, as assessed by quantification analysis. Using micro-autoradiography, an assessment of hybridized probe can be obtained by the silver grain count per cell or other unit area (Young and Kuhar, 1986; Shivers *et al.*, 1986b; McCabe *et al.,* 1986). When single cell resolution is not required, densitometric analysis can be carried out on autoradiography film exposed to sections, which are compared to calibrated radioactive standards (Steel *et al.*, 1989; McCafferty *et al.*, 1989).

In order to achieve reproducible results which can be reliably quantified, it is essential to standardize section thickness, target nucleic acid retention, hybridization and autoradiography conditions. It is also important to determine a standard curve, in order to correlate the measured grains or density to the level of nucleic acid (Davenport and Nunez, 1989; McCabe *et al.*, 1986).

In order to quantify the variation of β-TSH mRNA synthesis in the pituitary of euthyroid and hypothyroid rat, densitometric analysis was carried out on autoradiography film (Steel *et al.*, 1989; Figure 6). The probes were labelled with ^{35}S, as the labelling densities could be compared to ^{14}C standards, which are commercially available. The grey levels of the image were measured by computerized image analysis, and automatically converted into concentration of radioactivity. This method allowed the precise measurement of specific labelling for β-TSH in the pituitaries, which was shown to be 22 times greater in hypothyroid rats than in controls.

2.5 Tumour pathology

Endocrine tumours can be recognized by their histological appearance and the presence of secretory granules as revealed by ultrastructural investigation. Their characterization by immunocytochemistry has been widely studied (Polak and Van Noorden, 1986). However, it is not always possible to demonstrate immunoreactivity for a given peptide in tumour cells, because of peptide hypersecretion and limited storage capacity of the cells.

Such a tumour is the small cell carcinoma of the lung, which most commonly secretes gastrin-releasing peptide (GRP), a peptide thought to act as a growth factor in normal and neoplastic lung. The tumour cells are often poorly granulated, resulting in difficulty in demonstrating GRP-immunoreactivity in this tumour. However, *in situ* hybridization on both tissue sections and tissue culture preparation of this tumour has demonstrated that GRP gene expression is very high, and a strong hybridization signal was observed in discrete groups of cells throughout the tumour (Hamid *et al.*, 1989). Similarly, it was possible to provide a better characterization of gut carcinoid tumours by demonstrating the presence of beta-prepro-tachykinin mRNA in conjunction with immunoreactivity for substance P, neurokinin A and serotonin (Bishop *et al.*, 1989).

3. CONCLUSIONS

In all these examples *in situ* hybridization can offer a suitable complement or alternative to morphological investigation at light microscopical level by immunocytochemistry, and allows further insight into intracellular metabolic processes.

REFERENCES

Angerer RC, Cox KH and Angerer LM (1985) *In situ* hybridization to cellular RNAs. Genet. Eng. **7**: 43–65.

Bhatt B, Burns J, Flannery D and McGee JO'D (1988) Direct visualisation of single copy genes on banded metaphase chromosome by nonisotopic *in situ* hybridization. Nuc. Acids Res. **16**: 39513961.

Bishop AE, Hamid QA, Adams C, Bretherton-Watt D, Jones PM, Denny P, Stamp GWM, Hurt RL, Grimelius L, Harmar AJ, Valentino K, Cedermark B, Legon S, Ghatei MA, Bloom SR and Polak JM (1989) Expression of tachykinins by ileal and lung carcinoid tumours assessed by combined *in situ* hybridization, immunocytochemistry and radioimmunoassay. Cancer **63**: 1129–1137.

Brahic M, Haase AT and Cash E (1984) Simultaneous *in situ* detection of viral RNA and antigens. Proc. Natl. Acad. Sci. USA **81**: 5445–5448.

Brysch W, Hagendorff G and Schlingensiepen (1988) RNA probes transcribed from synthetic DNA for *in situ* hybridization. Nuc. Acids Res. **16**: 2333.

Casey J and Davidson N (1977) Rates of formation and thermal stabilities of RNA:RNA and DNA:DNA duplexes at high concentration of formamide. Nuc. Acids Res. **4**: 1539–1552.

Chan-Palay V, Yasargil G, Hamid Q, Polak JM and Palay SL (1988) Simultaneous demonstration of neuropeptide Y gene expression and peptide storage in single neurons of the human brain. Proc. Natl. Acad. Sci. USA **85**: 3213–3215.

Cox KH, De Leon DV, Angerer LM and Angerer RC (1984) Detection of mRNA's in sea urchin embryos by *in situ* hybridization using asymmetric RNA probes. Dev. Biol. **101**: 485–502.

Davenport AP and Nunez DJ (1990) Quantification in *in situ* hybridization. In: *In Situ Hybridization – Principles and Practice*. (Polak JM and McGee JO'D eds), Oxford University Press.

Denny P, Hamid Q, Krause JE, Polak JM and Legon S (1988) Oligoriboprobes: tools for *in situ* hybridization. Histochemistry **84**: 481–483.

Giaid A, Hamid H, Adams C, Springall DR, Terenghi G, Polak JM (1989a) Non-isotopic RNA probes. Comparison between different labels and detection systems. Histochemistry **93**: 191–196.

Giaid A, Gibson SJ, Ibrahim NBN, Legon S, Bloom SR, Yanagisawa M, Masaki T, Varndell IM and Polak JM (1989b) Endothelin 1, an endothelium-derived peptide is expressed in neurons of the human spinal cord and dorsal root ganglia. Proc. Natl. Acad. Sci. USA **86**: 7634–7638.

Gibson SJ, Polak JM, Bloom SR, Sabate IM, Mulderry PK, Morrison JFB, Kelly JS, Rosenfeld MG and Evans R (1984) Calcitonin gene-related peptide immunoreactivity in the spinal cord of man and eight mammalian species. J. Neurosci. **4**: 3101–3111.

Gibson SJ, Polak JM, Giaid A, Hamid QA, Kar S, Jones PM, Denny P, Legon S, Amara SG, Craig RK, Bloom SR, Penketh RJA, Rodek C, Ibraim NBN and Dawson A (1988) Calcitonin gene-related peptide mRNA is expressed in sensory neurons of the dorsal root ganglia and also in spinal motoneurones in man and rat. Neurosci. Lett. **91**: 283–288.

Gillam IC (1987) Non-radioactive probes for specific DNA sequences. Tib. Tech. **5**: 332–334.

Hamid Q, Wharton J, Terenghi G, Hassall CJ, Aimi J, Taylor KM, Nakazato H, Dixon JE,

Burnstock G and Polak JM (1987) Localization of atrial natriuretic peptide mRNA and immunoreactivity in the rat heart and human atrial appendage. Proc. Natl. Acad. Sci. USA **84**: 6760–6764.

Hamid QA, Bishop AE, Springall DR, Adams C, Giaid A, Denny P, Ghatei M, Legon S, Cutitta F, Rode J, Spindel E, Bloom SR and Polak JM (1989) Detection of human probombesin mRNA in neuroendocrine (small cell) carcinoma of the lung. Cancer **63**: 266–271.

Hofler H, Putz B, Rurhi C, Wirnsberger G, Klimpfinger M and Smolle J (1987) Simultaneous localization of calcitonin mRNA and peptide in a medullary thyroid carcinoma. Virchow. Arch. B **54**: 144–151.

Hopman AHN, Wiegant J, Raap AK, Landegent JE, van der Ploeg M and van Dujin P (1986) Bi-colour detection of two target DNAs by non-radioactive *in situ* hybridization. Histochemistry **85**: 1–4.

Inoue A, Yanagisawa M, Kimura 5, Kaswya Y, Miyauchi T, Goto T and Masaki T (1989) The human endothelin family: three structurally and pharmacologically distinct isopeptides predicted by three separate genes. Proc. Natl. Acad. Sci. USA **86**: 2863–2867.

Lawrence JB, Villnave CA and Singer RH (1988) Sensitive, high resolution chromatin and chromosome mapping *in situ*: presence and orientation of two closely integrated copies of EBV in lymphoma line. Cell **52**: 51–61.

Leuhersen KR and Baum MP (1987) *In vitro* synthesis of biotinylated RNA probes from A-T rich templates: problems and solutions. BioTech. **5**: 660–670.

McCabe JT, Morrell JI and Pfaff DW (1986) *In situ* hybridization as a quantitative autoradiographic method. In: *In Situ Hybridization in Brain*. (Uhl GR ed.), pp. 73–96, Plenum Press, New York.

McCafferty J, Cresswell L, Alldus C, Terenghi G and Fallon R (1989) A shortened protocol for *in situ* hybridization to mRNA using radiolabelled RNA probes. Techniques **1**:171–182.

Melton D, Kneg P, Rebagliati M, Maniatis T, Zinn K and Green MR (1984) Efficient *in vitro* synthesis of biologically active RNA and RNA hybridization probes from plasmid containing a bacteriophage SP6 promoter. Nuc. Acids Res. **12**: 7035–7050.

Mitchell AR, Ambros P, Gosden JR, Morten JEN and Porteous DJ (1986) Gene mapping and physical arrangement of human chromatin in transformed hybrid cells: fluorescent and autoradiographic *in situ* hybridization compared. Somatic Cell Mol. Genet. **12**: 313324.

Niedobitek G, Finn T, Herbst H, Bornhoft G, Gerdes J and Stein H (1988) Detection of viral DNA by *in situ* hybridization using bromodeoxyuridine-labelled DNA probes. Am. J. Pathol. **131**: 1–4.

Nunez W, Davenport AP, Emson PC and Brown MJ (1989) A quantitative *in situ* hybridization method using computer-assisted image analysis. Biochem J. **262**: 121–127.

Polak JM and Van Noorden S (1986) In: *Immunocytochemistry – Modern Methods and Applications*. 2nd edn. (Polak JM and Van Noorden S eds), Wright, Bristol.

Polak JM (1989) Regulatory peptides, Birkhausen Verlag, Basel.

Shivers BD, Harland RE, Pfaff DW and Schachter BS (1986a) Combination of immunocytochemistry and *in situ* hybridization in the same tissue section of rat pituitary. J. Histochem. Cytochem. **34**: 39–43.

Shivers BD, Harlan RE, Romano GJ, Howells RD and Pfaff DW (1986b) Cellular localization and regulation of proenkephalin mRNA in rat brain. In: *In Situ Hybridization in Brain*. (Uhl GR ed.), pp. 3–20, Plenum Press, New York.

Singer RH, Lawrence JB and Villnave C (1986) Optimisation of *in situ* hybridization using isotopic and non-isotopic detection methods. BioTechniques **4**: 230–250.

Springall DR, Bhatnagar M, Wharton J, Hamid Q, Gulbenkian S, Hedges M, Meleagros L, Bloom SR and Polak JM (1988) Expression of the atrial natriuretic peptide gene in the cardiac muscle of rat extrapulmonary and intrapulmonary veins. Thorax **43**: 44–52.

Steel JH, Hamid Q, Van Noorden S, Jones P, Denny P, Burrin J, Legon S, Bloom SR and Polak JM (1988a) Combined use of *in situ* hybridization and immunocytochemistry for the investigation of prolactin gene expression in immature, pubertal, pregnant, lactating and ovarectomized rats. Histochemistry **89**: 75–80.

Steel JH, Hamid Q, Van Noorden S, Chandrachud L, Jones P, Denny P, Burrin J, McNicol AM, Legon S, Bloom SR and Polak JM (1988b) Changes in prolactin and pro-opiomelanocortin messenger RNA in rat pituitary as shown by *in situ* hybridization. In: *Neuroendocrine Perspectives*, Vol. 6, (Scarlon MF, Wass JAH eds), Springer-Verlag, New York.

Steel JH, O'Halloran DJ, Jones PM, Chin WW, Bloom SR and Polak JM (1989) Simultaneous immunocytochemistry and *in situ* hybridization of beta thyroid-stimulating hormone and its messenger ribonucleic acid in euthyroid and hypothyroid rat pituitary. Mol. Cell. Probes, 4:385–396s.

Terenghi G, Polak JM, Ghatei MA, Mulderry PK, Butler JM, Unger WG and Bloom SR (1985) Distribution and origin of calcitonin generelated peptide immunoreactivity in the sensory innervation of the mammalian eye. J. Comp. Neurol. **233**: 506–516.

Terenghi G, Polak JM, Hamid Q, O'Brien E, Denny P, Legon S, Dixon J, Minth CD, Palay SL, Yasargil G and Chan-Palay V (1987) Localization of neuropeptide Y mRNA in neurons of human cerebral cortex by means of *in situ* hybridization with a complementary RNA probe. Proc. Natl. Acad. Sci. USA 84: 7315–7318.

Terenghi G and Fallon RA (1990) Techniques and applications of *in situ* hybridization. In: *Current Topics in Pathology – Pathology of the Nucleus*, Vol. 82, (Underwood JCE ed.), pp. 289–337, Springer-Verlag, Amsterdam.

Van der Ploeg M, Landegent JE, Hopman HHN and Raap AK (1986) Nonautoradiographic hybridocytochemistry. J. Histochem. Cytochem. **34**: 126–133.

Varndell IM, Polak JM, Sikri KL, Minth CD, Bloom SR and Dixon JE (1984) Visualisation of mRNA directing peptide synthesis by *in situ* hybridization using a novel single stranded cDNA probe: potential for the investigation of gene expression and endocrine cell activity. Histochemistry **81**: 597–601.

Wertmur JG, Ruyecham WT and Donthart RJ (1981) Denaturation and renaturation of Penicillin crysogenum mycophage double-stranded ribonucleic acid in tetraalkylammonium salt solution. Biochemistry **20**: 2999–3002.

Wolf S, Quaas R, Hahu U and Witting B (1987) Synthesis of highly radioactively labelled RNA hybridization probes from synthetic single-stranded DNA oligonucleotides. Nuc. Acids Res. **15**: 858.

Wolfson B, Manning RW, David LG, Arentzen R and Baldino F Jr (1985) Colocalization of corticotropin releasing factor and vasopressin mRNA in neurons after adrenalectomy. Nature **315**: 59–63.

Young WS and Kuhar MJ (1986) Quantitative *in situ* hybridization and determination of mRNA content. In: *In Situ Hybridization in Brain*. (Uhl GR ed.), pp. 243–248, Plenum Press, New York.

3
Application of *in situ* hybridization to studies of messenger RNA in the nervous system: functional correlations and their quantitation

J. de Belleroche, L. Virgo, A. Rashid and Y. Collaço Moraes
Department of Biochemistry, Charing Cross and Westminster Medical School, Fulham Palace Road, London W6 8RF

1. INTRODUCTION

With a tissue as heterogeneous and complex in its organization as the nervous system, *in situ* hybridization offers an approach of considerable potential. The complexity of this tissue means that a technique with high resolving power is needed to identify changes in gene expression that occur in a small group of cells. The CNS is extremely heterogeneous, it contains a dozen classical transmitters and more than 40 neuropeptides all in discrete cell clusters. This gives rise to a diversity of functional cell types each with its own strict localization and interaction with other cells. Activation of specific pathways may then result in only highly localized changes in metabolism which would be difficult to detect with tissue homogenates where the transmitter being studied may only be present in 1% of cells or less. The technique of *in situ* hybridization however allows the detection of highly localized changes in mRNA levels in specific cell types. The applications are numerous, from tracing pathways, localization of function to particular cell groups and localization of mRNAs of unknown function. Further, abnormal gene expression in neurological disorders may give some insight into the pathogenesis of disease and examples of these different applications will be given in the text. To satisfy the requirements for *in situ* hybridization it is essential to preserve the tissue during preparative treatments in order to be able to accurately define the cell types containing the mRNA being studied. Many of the mRNA species of particular interest are often far from abundant e.g. receptors and neuropeptides. Thus in studies of the nervous system, resolution, structural integrity and quantitation are key issues and these aspects will be discussed in detail together with the particular applications for which these approaches have been used.

1.1 Assay of mRNA as an index of functional activity

The rates of turnover of proteins are known to change in response to tissue requirements. This is largely regulated at the level of transcription and hence in these cases the level of mRNA would reflect functional activity, although other possible sites of regulation may also occur by control of peptide degradation. Measurement of changes in the level of a specific mRNA localize the site and define the mechanism of response to a perturbation. There are many examples where mRNA levels reflect physiological adaptations to such perturbations, one of the first of which to be demonstrated was for the neuropeptide vasopressin which plays a key role in homeostasis, regulating fluid balance and blood pressure. Levels of vasopressin increase several fold under conditions such as salt loading and dehydration and are depleted in Brattleboro rats which suffer diabetes insipidus. The large increase in peptide levels has been shown to be associated with a large increase in vasopressin mRNA in the magnocellular cells of the supraoptic nucleus and paraventricular nucleus of the hypothalamus (Uhl *et al.,* 1985). In Brattleboro rats a base change occurs in the DNA resulting in an abnormal mRNA and hence absence of the peptide. On the other hand, adrenalectomy which elevates plasma vasopressin, causes an increase in the vasopressin mRNA of a separate population of vasopressin containing cells in the hypothalamus, the parvocellular cells. Thus, *in situ* hybridization clearly defines the separate location of these functional changes. Furthermore, the vasopressin containing parvocellular cells also contain corticotropin releasing factor (CRF), which is released in the median eminence and controls the release of adrenocorticotropic hormone (ACTH) from the anterior lobe of the pituitary. As might be expected, adrenalectomy also causes an increase in CRFmRNA and this occurs in the same cells that show increases in vasopressin mRNA. The establishment of these results using conventional Northern and slot blot analysis would require the use of a much greater number of animals and would not give the clear localization that is possible with *in situ* hybridization. Thus maintenance of homeostasis is highly dependent on the rapid and sensitive regulation of transcription which is further reflected by concentrations of the peptide product. Not all changes in levels of mRNA may be as substantial and hence quantitation of levels of mRNA may be critical. The emphasis of this chapter will be on aspects of *in situ* hybridization which are particularly important to applications to studies of the nervous system, such as resolution of changes in mRNA, their quantitation and applications to the study of neurological disease and the special problems associated with handling autopsy material.

2. STANDARD PROCEDURES THAT ARE APPROPRIATE FOR HANDLING CNS TISSUE

2.1 Tissue preparation, permeabilization and prehybridization

A similar protocol for tissue preparation can be used for most applications to the nervous system. Frozen sections (4–15 μm) are most widely used. These are thaw mounted onto subbed slides, fixed in paraformaldehyde and taken through a number of permeabilization stages to increase the penetration of probe into the tissue. Tissue is then incubated in a prehybridization buffer to block non-specific binding sites. Charge interactions between positively charged proteins and negatively charged probes can give rise to artefactual signals. Acetylation is carried out to mask positive charges. Permeabilization can consist of a number of steps in which tissue is exposed to 0.1 M HCl, proteases, triton and lipid solvents such as chloroform. Prehybridization is carried out in a saline medium (phosphate buffer containing 2 × SSC :[3 M sodium chloride, 0.3 M sodium citrate]) containing sheared single stranded DNA, pyrophosphate, albumin, ficoll and polyvinylpyrrolidone to reduce non-specific binding sites. Dithiothreitol may be used to reduce non-specific binding when ^{35}S probes are being used and dextran sulphate may also be useful to accelerate the rate of association of probe and hence effectively increasing probe concentration.

In situ hybridization has been demonstrated in formalin-fixed paraffin-embedded tissues which greatly enhances the flexibility and versatility of this technique. This approach has a large number of advantages in that improved morphological detail is obtained and storage and processing of material is facilitated. In addition large reserves of routinely fixed human autopsy material may be accessed (Moench, 1987). However, optimization of signal can prove difficult with some probes.

2.2 Choice of probes and isotopes

A range of cDNA, cRNA and oligonucleotide probes have been successfully applied to *in situ* hybridization and each type of probe has its own advantages. The most widely used probes are cDNA probes which are easy to use and can be obtained at high specific activity but they have the disadvantage that they readily reanneal in solution. Oligonucleotide probes are more permeable and can be tailor-made but produce less stable hybrids and may show lower specificity. cRNA probes have the advantages that they form stable hybrids, can be synthesized to high specific activity and are single stranded but have the disadvantage that they often exhibit probe "stickiness". A further advantage of cRNA probes is that they can readily be synthesized with an SP6 promoter in both sense and anti-sense orientations to provide both a control probe and an active probe. The most commonly used isotopes are ^{32}P, ^{35}S and ^{3}H and the choice of isotope will depend on the application. Quantitation can only be carried out reliably using the lower energy emitters, ^{35}S and ^{3}H and subcellular localization can only be carried out using tritium due to the short path length of emission. Although not used for either of these applications, ^{32}P is in fact quite widely used for initial rapid screening experiments where optimal conditions of hybridization and washing are being established and also where mRNA concentrations are low.

2.3 Hybridization

Hybridization is carried out in the prehybridization buffer containing the labelled probe at an appropriate concentration for the application required. It is useful to include formamide in the buffer to reduce the temperature of hybridization and subsequent washes will need to be optimized for each probe. For cDNA probes, a temperature of 40°C gives good results in the presence of formamide. Higher temperatures tend to be deleterious for the tissue.

2.4 Detection

It is important to ensure that optimal conditions of development are used to obtain the best signal due to specific hybridization with a low background. Best results are obtained when emulsions are developed (Kodak D19) for up to 5 min. Development of silver grains in the background will occur if development is carried out for periods greater than 7 min.

2.5 Resolution

There are three levels of resolution which have proved useful in studies of the nervous system:

(i) Autoradiography using X-ray or tritium-sensitive film. This gives good regional resolution and is widely used for a preliminary survey (Figure 1).

(ii) Emulsion coated slides (or coverslips) viewed darkfield. This allows clear localization of signal to cell groups and is generally used at a magnification of approximately ×50. Darkfield illumination facilitates the localization of silver grains which are seen as white spots against a black background.

(iii) Emulsion coated slides counterstained and viewed at high magnification to localize hybridization signals in specific cells. Magnification of ×500 would be commonly used for this purpose. Quantitation may be carried out by grain counting or by densitometric analysis of tissue sections. In densitometry the tissue sections are compared to radioactive standards, opposed to emulsion and exposed in parallel.

Nuclear emulsions are mainly used in the form of a liquid emulsion such as NTB-2 (grain diameter 0.26μm) which detects β-emitters. The slides bearing the tissue sections are dipped into diluted emulsion (e.g. diluted 1:1 with 1% glycerol solution) and allowed to dry, draining vertically in a dark dry, atmosphere. Alternatively, emulsion coated gelatin strips (Kodak stripping film) may be used where a thin layer (5μ) of emulsion is deposited on gelatin. The latter preparation has the advantage that a uniform layer of emulsion is formed but the presence of the gelatin may interfere with subsequent staining procedures.

2.6 Controls

It is always important to authenticate the signal that is detected. Initially, the signal should be shown to be sensitive to treatment with RNase (Figure 2), the distribution should be consistent with that of the protein encoded by the mRNA and thirdly mRNA isolated from the tissue should have the same hybridization characteristics. An example of this type of control is shown in the authentication of the distribution of cholecystokinin (CCK) mRNA in mammalian brain (Figure 3). The RNase sensitivity is tested by pretreatment of sections with RNase A (100μg/ml in 0.5 M NaCl, 10 mM TrisHCl, pH 8.0, 1 mM ethylenediamine tetra acetic acid) for 30 min at 37°C prior to hybridization with the radioactive probe. This results in the loss of label in the CCK-containing region of cerebral cortex, hippocampus and thalamus. Cells in this region are known by immunohistochemistry to contain CCK peptide and tissue homogenates from these regions are known to be rich in CCK-immunoreactivity. A further type of control that can be carried out is to selectively lesion cells containing the product in order to show the loss of the mRNA in parallel and this has been done by lesion of the claustrum (which give rise to the innervation of caudate nucleus). This lesion results in the loss of CCK-mRNA from the claustrum. Hybridization with a 20-fold excess of unlabelled CCK-cDNA probe prior to hybridization also results in a loss in hybridization signal.

Unexpected homologies can give rise to misleading results and it may prove

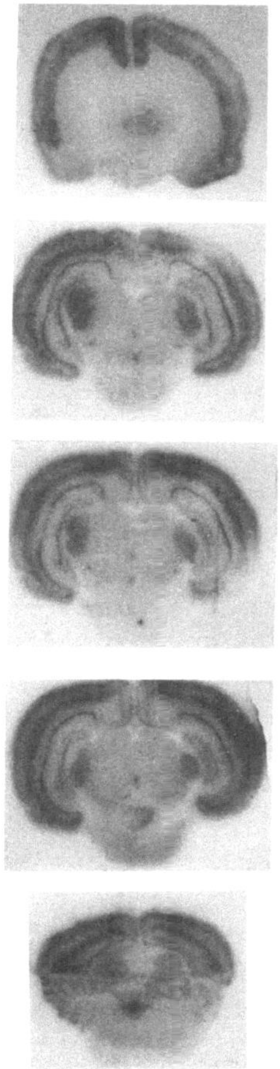

Figure 1 Regional distribution of CCKmRNA in rat brain determined by autoradiography. Thin sections of rat brain were hybridized for 16 h with ^{32}P-CCK-cDNA. A distinct pattern of distribution is shown with high concentrations of CCK mRNA being present in cerebral cortex, hippocampus, thalamus, inferior colliculus and periaqueductal grey matter whilst low levels are found in caudate nucleus. Data reproduced from de Belleroche *et al.* (1990) with acknowledgement to Churchill Livingstone.

necessary to use an adjacent section of probe sequence. In addition, if the probe contains GC-rich segments, non-specific hybridization may occur which can be avoided by use of a sub-segment of the probe which does not contain this section. In some cases chemography can occur in which there is an interaction between components of the section and the photographic emulsion which can give a convincing signal. It is important to control for this by exposure of slides to emulsion without application of the radioactively labelled

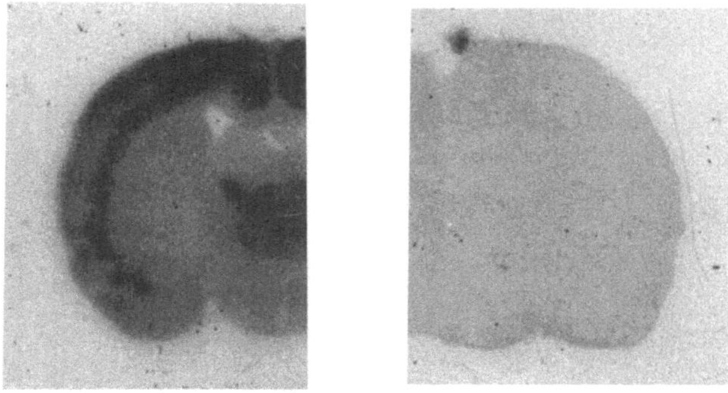

Figure 2 Distribution of CCKmRNA in rat brain and its sensitivity to RNase. Thin sections of rat brain were hybridized for 16 h with ^{32}P-CCK-cDNA. In the right hand section, pretreatment with RNase (100 μg/ml) was carried out prior to hybridization. The RNase sensitive signal is shown to be present in cerebral cortex and thalamus but absent from caudate nucleus. Data reproduced from de Belleroche *et al.* (1990) with acknowledgement to Churchill Livingstone.

probe. If this proves to be the case the specificity of signal should be tested out as described above i.e. RNase treatment, pretreatment with excess unlabelled probe to test for specificity and hybridization with a probe containing an adjacent sequence. A generalized Nissl-like pattern of hybridization may be due to non-specific hybridization which correlates with the regional density of RNA and is high in regions such as the granule cell layer of cerebellum and the dentate and CA 1–4 fields of the hippocampus. This can usually be overcome by improving the stringency of washing and by increasing the time and ionic strength of washes. A myelin-like pattern may be obtained by interactions with lipids and this can be overcome by treatment with lipid solvents such as acetone or chloroform.

3. QUANTITATIVE AUTORADIOGRAPHY

Autoradiographic images can be quantitated densitometrically by comparison to tissue standards containing radioactive standards. The standards are made by mixing fixed amounts of radioactivity such as ^{35}S- or ^{3}H-dCTP with portions of a brain paste. A suitable range of concentrations to use for ^{35}S standards would be 100–6000 dpm/mg tissue. Blocks of these brain paste standards are frozen and used for making frozen sections. The radioactive standard sections are then treated in the same way as tissue sections and exposed to emulsion with each film or batch of slides coated with emulsion. Quantitation requires the use of a computerized image analyser. For this, a video camera is usually used to obtain the image of the autoradiogram, which is digitized to yield a grey value between 0 and 256 for each pixel in the image. Variation in

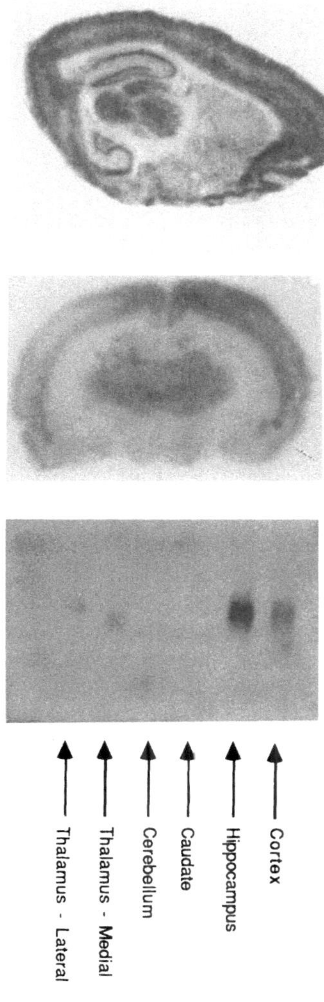

Figure 3 Distribution of CCKmRNA in rat brain: comparison of distribution obtained by *in situ* hybridization with that obtained by Northern blot analysis. RNA was extracted from various brain regions, size fractionated on agarose, blotted onto nylon filters, hybridized with ^{32}P-CCK-cDNA and the distribution compared with that obtained by *in situ* hybridization. Relative levels of hybridization signal were obtained by both methods, with high concentrations of CCKmRNA in cerebral cortex and hippocampus, moderate levels in thalamus and low levels in caudate. Data reproduced from de Belleroche *et al* (1990) with acknowledgement to Churchill Livingstone.

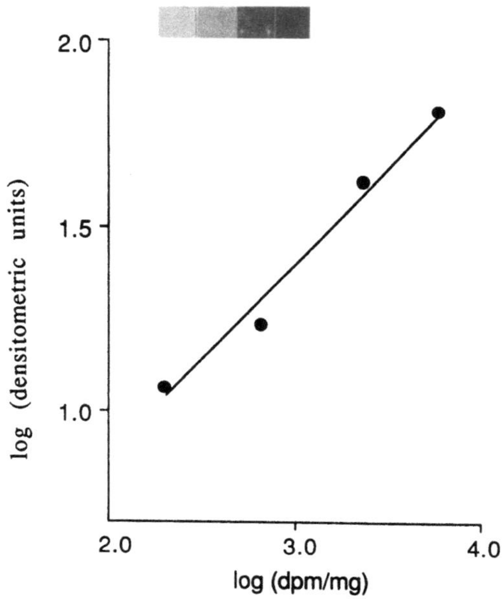

Figure 4 Dose response curve to ^{35}S-radioactive brain paste standards. Four brain paste standards were sectioned and apposed in parallel with brain sections. The densitometric values were sampled and a plot of log densitometric units against log (dpm/mg) is shown to be linear. The inset shows the photographic images from which these values were obtained. Data reproduced from de Belleroche *et al.* (1990) with acknowledgement to Churchill Livingstone.

illumination is controlled for by a shading correction. The signals obtained with the brain paste standards are used to construct a dose response curve (Figure 4). Mean values of dose (dpm/mg) in each region of interest are then derived from adjacent RNase and untreated sections. The RNase-sensitive signal is then obtained by subtraction. Multiple sampling from each area of interest is needed for each section.

For determining tissue concentrations of mRNA, it is also important to ensure that concentrations of the probe are optimized for obtaining accurate information. From the dose response curve obtained from the hybridization signal, it can be seen that a sufficient concentration of probe must be used in order to saturate the tissue levels of mRNA present in the sample (Figure 5). If a lower concentration is used, then an underestimation of concentration will be obtained. Alternatively, relative estimates can be obtained under conditions where lower concentrations of probe are used.

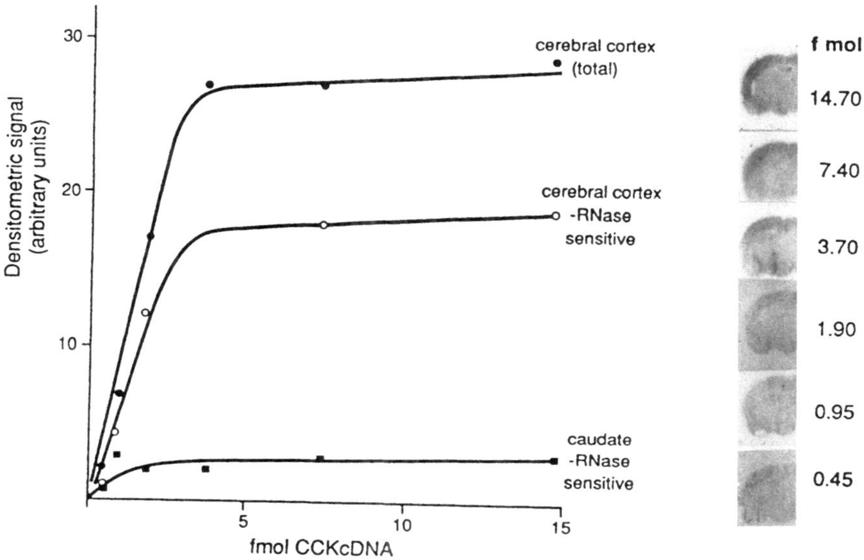

Figure 5 Saturation analysis of CCK-mRNA in rat brain. Sections of rat brain were hybridized with a range of concentrations of ^{32}P-CCK-cDNA (0.45–14.47 fmol/section) in the presence or absence of RNase (100 μg/ml) and the densitometric signal measured in cerebral cortex and caudate nucleus. The average value for the densitometric signal in the absence of RNase (total) is shown together with the RNase-sensitive signal obtained by subtraction of the values obtained in the presence of RNase from total values (RNase-sensitive) for both cerebral cortex and caudate nucleus. Data reproduced from de Belleroche *et al.* (1990) with acknowledgement to Churchill Livingstone.

4. APPLICATIONS

4.1 Localization

The earliest applications of *in situ* hybridization in the nervous system have been to localize specific mRNAs within specific neuronal cells and there are a number of examples of neuropeptide localization. This approach has been extended to look at variations in peptide mRNA concentration during development (de Belleroche *et al.,* 1990) and other physiological conditions (e.g. diurnal variations in vasopressin and oxytocin mRNA: Burbach *et al.,* 1988). More recently, greater emphasis has been placed on more specific functional correlates of mRNA levels and this is discussed in the following section.

4.2 Functional studies

There are now an increasing number of applications where it is possible to localize function to specific cell types. One particular area where this has been effectively applied has been in the investigation of responses to environmental challenges and drug treatment. For example, investigations of the neuronal control of endocrine responses to opiate withdrawal and stress have shown that these treatments are associated with large increases in enkephalin mRNA in the parvocellular cells of the paraventricular nuclei of the hypothalamus (Lightman and Scott Young III, 1987). Through their extensive innervation these cells are then likely to have profound effects on the release of luteinizing hormone releasing hormone (LHRH), corticotropin releasing factor (CRF), thyrotropin releasing hormone (TRH) and dopamine into the median eminence and hence regulate anterior pituitary function.

Specific changes in mRNA have also been localized in studies of responses to noxious stimuli, where it has been demonstrated that preprotachykinin-A mRNA is specifically increased in rat dorsal root ganglia (Noguchi *et al.,* 1988).

In models of neurodegenerative conditions, the rapid changes in gene expression that occur in response to injury can be localized and give insight into the mechanisms involved. One such response is the rapid induction of c-fos mRNA in cerebral cortex after excitotoxin injection (Wood and de Belleroche, 1991) which is demonstrated in Colour plate 2.

A number of novel sequences have been discovered in recent years through homologies with common sequences such as those present in members of the ion channel and G-protein-linked receptor families and *in situ* hybridization has proved invaluable in characterizing the localization of specific mRNAs and unravelling the complexity of receptor function. For example it has been shown that the GABA receptor alpha subunit mRNAs each show a distinct pattern of distribution and hence are closely associated with different functions (Wisden *et al.,* 1988,1989 Montpied *et al.,* 1988). Similar studies have been carried out on mRNA distribution of muscarinic receptor subtypes (Buckley *et al.,* 1988) and the 5-HT_{1C} receptor subtype (Hoffman and Mezey, 1989). The localization of components of transduction systems linked to specific receptors has also been explored by *in situ* hybridization and a similar degree of differential localization is seen in this area too, for example that shown by mRNA localization of Go mRNA which is a particularly abundant G protein in the brain (Colour plate 3) and phospholipase C isozymes (Ross *et al.,* 1989).

More recently, screening of cDNA libraries derived from mammalian brain with a substance K receptor probe has yielded a novel sequence of unknown function. Localization of the mRNA by *in situ* hybridization to specific brain regions has played an important role in uncovering the function of this mRNA (Matsuda *et al.*, 1990). It has turned out to represent a potential cannabinoid receptor which is likely to yield useful results about the establishment of addiction to tetrahydrocannabinol and related drugs.

4.3 Application of *in situ* hybridization to studies of neurological disorders

The technique of *in situ* hybridization has now a well proven record in applications to the study of experimental animals. However, there are a number of important applications where study of autopsy material is necessary to get an insight into the pathology associated with neurological and psychiatric disorders. In this case the length of delay between death and autopsy and the storage conditions used can have serious effects on the levels of mRNA. It is however possible to obtain a good signal from material taken within 24 h of death. We have used this approach to study the pathogenesis of motor neurone disease where there is a selective loss of motor neurones in the spinal cord, brain stem and motor cortex. A clear-cut localization of choline acetyl transferase (ChAT) mRNA, which is a marker of all cholinergic cells can be seen in control sections of spinal cord compared to tissue obtained from MND cases (Figure 6). Further, the signal can be shown under higher magnification to be localized to the anterior horn cell region (Figure 7). This information can then be used as a reference point to which changes in surrounding neuronal and glial cells can be related.

Another example of the application of *in situ* hybridization to neurological disease is seen in studies of the expression of the β-amyloid gene in Alzheimer's disease. High concentrations of β-amyloid in senile plaques are one of the hallmarks of this condition and although β-amyloid is present throughout the body, an increased expression is indicated in Alzheimer's disease. Results with *in situ* hybridization indicate that an increase in the expression of an alternatively spliced mRNA which lacks the sequence coding for the Kunitz protease inhibitor may be important in the pathology of Alzheimer's disease (Palmert *et al.,* 1988; Schmechel *et al.,* 1988).

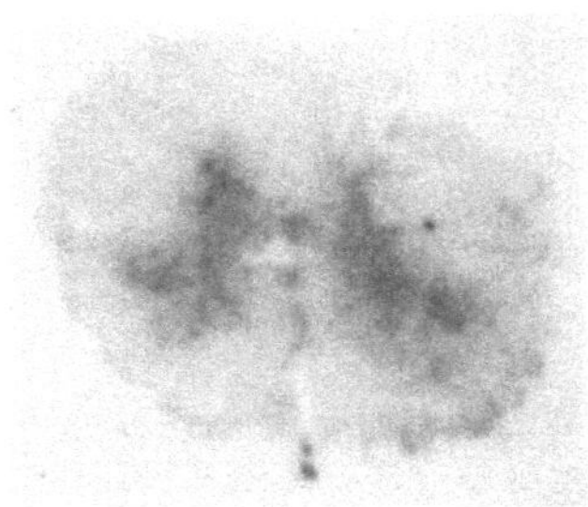

Figure 6 Distribution of choline acetyltransferase in human spinal cord. Thin sections of control human lumbar spinal cord were hybridized with ^{32}P-labelled ChAT-oligonucleotide probe. High concentrations of ChAT-mRNA are seen in dorsal and ventral grey matter, being especially concentrated in the region of the anterior horn cells. This distribution corresponds to that of the ChAT enzyme containing cells. Data is taken from Virgo *et al.* (1992).

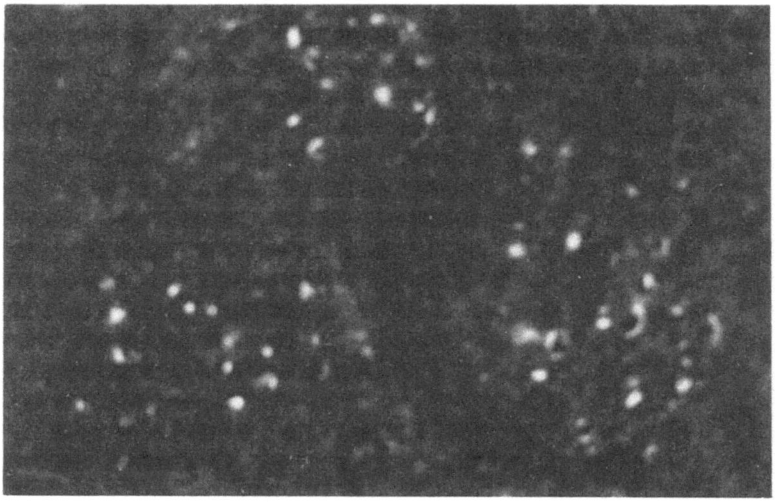

Figure 7 Darkfield view of anterior horn cells in human lumbar spinal cord. Thin sections of human spinal cord were hybridized as described in Figure 8 and viewed darkfied (×200). The signal is localized in clusters coinciding with the distribution of anterior horn cells. Data is taken from Virgo *et al.* (1992).

5. CONCLUDING REMARKS

Current research indicates the enormous complexity of the nervous system with the emergence of large numbers of neurotransmitters, their receptor subtypes and the associated transduction systems. Techniques which allow the fine localization of these components have been important in characterizing these systems. There is little doubt that this approach will continue to provide useful information in these areas and in further applications to understand the pathogenesis of neurological disease.

ACKNOWLEDGEMENTS

We are grateful to the Wellcome Trust and the MNDA for financial support towards the work discussed in this chapter.

REFERENCES

Buckley NJ, Bonner TI and Brann MR (1988) Localization of a family of muscarinic receptor mRNAs in rat brain. J.Neurochem. **8**: 4646–4652.

Burbach JP, Liu B., Voorhuis TA and Van Tol HH (1988). Diurnal variation in vasopressin and oxytocin messenger RNA in hypothalamic nuclei of the rat. Brain Res. **464**: 157–160.

de Belleroche J, Bandopadhyay R, King A, Rashid A and Malcolm ADBM. (1990) Regional distribution of cholecystokinin messenger RNA in rat brain during development: quantitation and correlation with cholecystokinin immunoreactivity. Neuropeptides **15**: 201–212.

Hoffman B and Mezey E (1989) Distribution of serotonin $5HT_{1c}$ receptor mRNA in adult rat brain. FEBS Lett, **247**: 453–462.

Lightman SL.and Scott Young IIIW (1987) Changes in hypothalamic preproenkephalin A mRNA following stress and opiate withdrawal. Nature **326**: 643–645.

Matsuda LA, Lolait ST., Brownstein MJ, Young AC. and Bonner TJ. (1990) Structure of a cannabinoid receptor and functional expression of the cloned cDNA. Nature **436**: 561–564.

Moench TR (1987) *In situ* hybridization. Molecular and Cellular Probes **1**: 195–205.

Montpied P, Martin BM, Cottingham SL, Stubblefield BK, Ginns EI and Paul SM (1988) The regional distribution of the $GABA_A$/benzodiazepine receptor (alpha subunit) mRNA in rat brain. J. Neurochem. **51**: 1651–1654.

Noguchi K, Morita Y, Kiyama H, Ono K and Tohyama M (1988) A noxious stimulus induces the preprotachykinin-A gene expression in rat dorsal root ganglion: a quantitative study using *in situ* hybridization histochemistry. Brain Res. **464**: 31–35.

Palmert MR, Golde TE, Cohen ML, Kovacs DM, Tanzi RE, Gusella JP, Usiak MF, Younkin LH and Younkin SG. 1988. Amyloid protein precursor messenger RNAs : differential expression in Alzheimer's disease. Science **241**: 1080–1084.

Ross CA, Mac Cumber MW, Glatt CE and Snyder SH. (1989) Brain phospholipase C isozymes: differential mRNA localizations by *in situ* hybridization. Proc. Natl. Acad. Sci. USA **86**: 2923–2927.

Schmechel DE, Goldgaber D, Burkhart DS, Gilbert JR, Gajdusek DC and Roses AD (1988) Cellular localization of mRNA encoding amyloid beta protein in normal tissue and Alzheimer disease. Alzheimer Dis. Assoc. Disord. **2**: 96–111.

Uhl GR, Zingg HH and Habener JF (1985) Vasopressin mRNA *in situ hybridization*: Localization and regulation studied with oligonucleotide cDNA probes in normal and Brattleboro rat hypothalamus. Proc. Natl. Acad. Sci. USA **82**: 5555–5559.

Virgo L, de Belleroche J, Ross, M and Steiner TJ (1992) Characterisation of the distribution of choline acetyltransferase messenger RNA in human spinal cord and its depletion in motor neurone disease: submitted.

Wisden W, Moore BJ, Darlison MG, Hunt SP and Barnard EA. (1989) Localization of $GABA_A$ receptor α-subunit mRNAs in relation to receptor subtype. Mol Brain Res. **5**: 305–310.

Wisden W, Moore BJ, Darlison MG, Hunt SP and Barnard EA. (1988) Distinct a$GABA_A$ receptor alpha subunit mRNAs show differential pattern of expression in bovine brain. Neuron **1**: 937–947.

Wood H and de Belleroche J. (1991) Induction of c-fos mRNA in cerebral cortex by excitotoxin stimulation of cortical inputs: involvement of *N*-methyl-D-aspartate receptors. Brain Research. **545**: 183–190.

4
The use of *in situ* hybridization in studies of viral disease

Adrienne L. Morey and Kenneth A. Fleming
University of Oxford, Nuffield Department of Pathology and Bacteriology, John Radcliffe Hospital, Headington, Oxford 0X3 9DU

1. INTRODUCTION

In situ hybridization has a number of features which make it particularly suitable for use in studies of viral disease. Not only can it confirm the presence of specific viral DNA or RNA sequences in a range of histological preparations, but by demonstrating the precise tissue, cellular and subcellular location of the virus it can correlate the presence of a virus with its pathological effects and provide insight into the mechanisms involved in virus-host cell interactions. Such information is necessarily lost when the more conventional technique of dot (filter) hybridization of extracted nucleic acid, or the more recently described polymerase chain reaction (PCR) are used to detect viral sequences. As well as localizing viral genomes in the episomal or integrated state, *in situ* detection of viral mRNA is possible. This can provide valuable information about the level of viral gene expression and sites of viral protein synthesis. In addition, *in situ* hybridization can be combined sequentially with immunohistological labelling of either cellular antigens (to identify unequivocally the cell types infected) or viral antigens (to determine whether viral nucleic acid is being translated into protein products), thus increasing the amount of information available from a given sample of tissue.

Although the great potential of *in situ* hybridization as a diagnostic and investigative tool in studies of viral disease was evident early in the history of this technique (Orth *et al.*, 1970; McDougall *et al.*, 1972), it is only relatively recently that technological and methodological advances have made possible the realization of this potential. The cloning of genetic material from most important DNA and RNA viruses has enabled the production of almost unlimited quantities of a wide variety of probes; in addition, the availability of sequence data on these viruses has made possible the construction of highly specific oligonucleotide probes. The commercial availability of an increasing range of probes has put *in situ* hybridization within reach of non-specialized laboratories. The development of non-isotopic labelling and detection systems has led to dramatic improvements in the speed and resolution of the technique, while reducing the costs and hazards associated with radioactively labelled probes. Initial problems with reduced sensitivity using non-isotopic probes have been largely overcome by continuing methodological refinements. Tissue digestion techniques adapted from immunohistochemistry have made *in situ* hybridization possible on formalin-fixed, paraffin-embedded tissue, thus making vast archives of stored material with good tissue morphology available for retrospective studies. These advances have led to a broadening of the possible applications of *in situ* hybridization, and an exponential increase in the number of laboratories using the technique as a research or diagnostic tool (Fleming, 1987; Grody *et al.*, 1987; Hofler, 1987).

In order to highlight those areas in which *in situ* hybridization can provide unique insights into the process of viral infection, we have chosen to discuss the use of the technique in viral studies in terms of its particular strengths, rather than reviewing all possible applications. We have defined *in situ* hybridization as the use of labelled nucleic acid probes to detect complimentary sequences in tissue sections, cellular preparations, or chromosome spreads. Protocols involving the use of non-cellular preparations (such as filter and sandwich hybridizations) to detect viral genomes are not included, but have been described in previous reviews (Bornkamm *et al.*, 1983; Maitland *et al.*, 1987; Norval and Bingham, 1987). While we have focused on the use of *in situ* hybridization in the study of human viral diseases, it should be noted that the technique has been extensively employed in the study of viral infections in other species, both as animal models of human disease and in the context of veterinary medicine (e.g. Orth *et al.*, 1970; Brahic *et al.*, 1981; Brahic *et al.*, 1984; Stowring *et al.*, 1985; Peluso *et al.*, 1985; Gendelman *et al.*, 1986; Baskar *et al.*, 1986; Jilbert *et al.*, 1987; Alexandersen *et al.*, 1987; Perlman *et al.*, 1989; Lipkin *et al.*, 1989). The technique has also been employed in analysis of insect-borne viruses (Ballinger *et al.*, 1988) and plant viroids (Harders *et al.*, 1989).

Specific applications of *in situ* hybridization in the context of viral disease fall into three main categories:

(i) as a diagnostic tool where other methods are slow or inadequate,

(ii) to investigate virus-host cell interactions (including the role of viruses in oncogenesis),

(iii) in studies of basic viral biology, including mechanisms of replication.

Before reviewing these three categories, a number of general methodological points with relevance to viral studies should be made.

1.1 Methodological considerations relevant to viral *in situ* hybridization

1.1.1 Types of probe

The researcher aiming to become familiar with *in situ* hybridization techniques is faced with an increasing variety of different probe types, probe labelling strategies, hybridization protocols and detection systems (see Chapter 1, also reviews by Coghlan *et al.*, 1985; Moench, 1987; Maitland *et al.*, 1987; Grody *et al.*, 1987; Norval and Bingham, 1987; Hofler, 1987; Myerson, 1988). The optimal strategy for each particular application will vary depending on the origin and preparation of the tissue being probed, and the nature and location of the target nucleic acid.

The greatest published experience is with double-stranded DNA probes, labelled with either radioactive or non-radioactive reporter molecules by nick translation or random priming techniques. The use of stable non-isotopic labels such as biotin has made possible the commercial availability of labelled double-stranded DNA probes for several viruses (Enzo Biochem; Life Technologies; Euro-Diagnostics). The production of single-stranded RNA riboprobes using specialized plasmid vectors containing dual RNA polymerase promoter sequences has introduced a new degree of flexibility to *in situ* hybridization studies; sense and antisense probes can be used to determine the polarity of the target nucleic acid, and have been especially useful in investigations of viral gene expression in latent infections. Unfortunately biotinylated nucleotides are not efficient substrates for the SP6 RNA polymerase, and thus most workers have employed isotopic labels for riboprobes; the development of alternative non-isotopic labelling strategies will undoubtedly be of major importance in broadening the range of applications of such probes. Advances in the technology for synthesizing oligonucleotides have made possible the preparation of short (20–60 base pair), single-stranded, customized DNA probes at relatively low cost. Based on sequence data, they can be designed to be "unique" for a particular strain of virus, or to contain sequences shared by a family of viruses. While eliminating the need for purification and cloning of viral nucleic acid, such probes suffer from limited sensitivity because of the low level of incorporation of reporter molecules and the usual requirement for 3′ or 5′ "end labelling". Labelling with non-isotopic markers is possible, however the use of cocktails of different oligonucleotides is sometimes necessary in order to achieve satisfactory sensitivity. A final method of probe production employs the polymerase chain reaction (PCR). This technique employs a heat stable DNA polymerase from *Thermus aquati-*

cus (Taq polymerase) to amplify a chosen segment of DNA by many orders of magnitude. Labelled nucleotides (either isotopic or non-isotopic) can be directly incorporated into the amplified sequence, producing a highly specific probe with theoretically greater sensitivity than oligonucleotide probes. A technique for producing biotinylated hepatitis B virus probes using PCR has been developed in this laboratory (Lo *et al.*, 1988), and studies are currently underway to extend this technique to other non-isotopic labels such as digoxigenin, and to optimize PCR-labelled probe characteristics for use in *in situ* hybridization.

1.1.2 Probe labelling strategies

The choice of probe label depends on the particular requirements of the investigation. While high energy emission isotopic probes (labelled with ^{32}P or ^{125}I) are ideally suited to macroscopic studies of viral distribution in whole animal or organ sections, they give poor subcellular resolution compared to non-isotopic probes. Low energy isotopic probes (incorporating ^{3}H) provide good resolution, but require long periods for autoradiographic detection. Probes labelled with ^{35}S have recently gained popularity, providing a reasonable compromise between requirements for resolution, sensitivity and speed. An advantage of isotopic labelling is that it is possible, by grain counting, to quantitate the amount of bound probe. Recent methodological advances have improved the sensitivity of non-isotopic detection systems to provide comparable sensitivity with ^{35}S-labelled probes, however quantification of target nucleic acid using non-isotopic systems is problematic. Non-isotopic protocols are generally more rapid than isotopic protocols, and being free from radiation hazards are potentially more useful as diagnostic assays in the general laboratory setting. The range of non-isotopic labels available is steadily increasing (see Chapter 1). Hybrids between labelled probe and target can be detected by histochemical or immunohistochemical means and visualized using fluorescent conjugates, a variety of coloured enzyme substrates, or colloidal gold. While fluorescent detection is sensitive and affords good resolution, it is generally unsuitable for use on tissue sections because of tissue autofluorescence, and has the additional disadvantage of fading over time. Chromogenic substrates provide a more permanent means of detecting hybridization signal, and substrates of different colours can be chosen as appropriate to allow counterstaining or multiple labelling of more than one nucleic acid target (see below). Colloidal gold has only recently been employed for non-isotopic labelling of viral sequences in tissue sections (Cubie and Norval, 1989; Morey *et al.*, 1991a). It has the advantages of non-toxicity, highly localized signal and compatibility with standard haematoxylin/eosin staining (Figure 1).

If the viral genome under investigation is too large to be cloned in a single

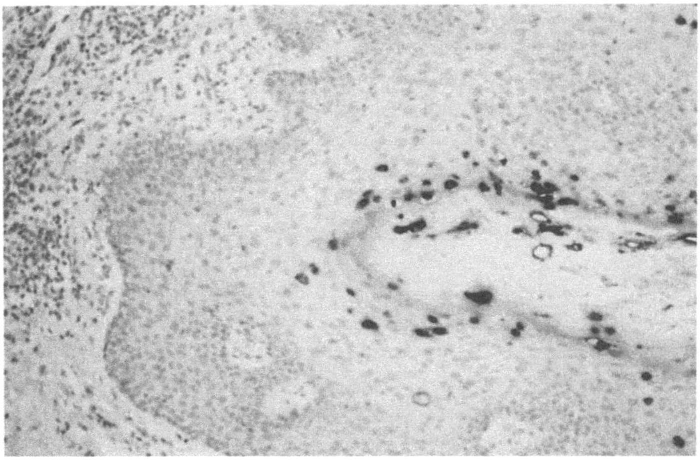

Figure 1 Formalin-fixed, paraffin embedded section of an anogenital condyloma shown to be positive for human papillomavirus type 6 using a digoxigenin labelled DNA probe detected by immunogold labelling with silver amplification. Haematoxylin and eosin counterstain. Infected cells in the upper layers of the epithelium appear black.

vector, probes to the different segments of the genome can be combined in order to maximize sensitivity. Unrelated viral probes can also be combined in a single hybridization reaction to screen for the presence of any one of the component viruses. If different reporter molecules are attached to different probes, simultaneous infection of a single cell by more than one virus can be detected. This was first shown using ^{3}H and ^{35}S probes to measles and visna viruses and a two-colour microradioautographic detection system (Haase *et al.*, 1985a), but has more recently been successfully used to detect human papillomavirus (HPV) and herpes simplex virus (HSV) in single cells using biotinolyted and haptenized probes with different enzymatic detection systems (Mullink *et al.*, 1989a). Viral and human genomic DNA sequences have also been simultaneously detected with biotin and digoxigenin labelled probes using different coloured enzymatic substrates (Herrington *et al.*, 1989; Morey *et al.*, 1991a). Double DNA target detection can also be combined with prior immunohistological labelling of specific antigens to give triple colour detection of three different targets in a single cell (Morey *et al.*, 1991a; Colour plate 4). Viral and human mRNA species have also been simultaneously detected using ^{35}S and biotin labelled probes (Ozden *et al.*, 1990). While the co-detection of viral genomes and viral or cellular mRNAs is a possibility, obligate differences in the *in situ* hybridization protocols for DNA and mRNA make such a combination currently suboptimal (unpublished observations).

1.1.3 Tissue digestion protocols

Viral sequences have successfully been detected in chromosome spreads, cell smears, frozen sections, and fixed tissues embedded in both paraffin and plastic. An important consideration in all fixed tissues is the requirement for proteolytic "digestion" to remove protein cross-linking and render the target nucleic acid available for hybridization. A range of different agents have been used with variable effectiveness (Brigati *et al.*, 1983; Burns *et al.*, 1987; Burns *et al.*, 1988; Naoumov *et al.*, 1988a). It is our experience that when testing archival clinical samples, digestion protocols must be optimized for each particular block of tissue, as the amount of proteolytic digestion required will depend not only on the type and duration of fixation, but also on the origin and state of the tissue prior to fixation. We have found that pretreatment of slides with the adhesive 3-aminopropyltriethoxysilane (Burns *et al.*, 1987) is crucial to ensure retention of sections during proteolytic digestion. In order to ensure that sufficient digestion has been performed and to provide an index of sensitivity, the use of an internal control is advisable. When probing routinely processed tissues for viral DNA, we also hybridize an adjacent section with probe pHY2.1 which detects a 2.12 kb sequence occurring approximately 2000 times on the long arm of the Y chromosome (Cooke *et al.*, 1982). Between 100 and 200 highly homologous 2.0 kb sequences are also present elsewhere in the genome. Unless extremely high stringency conditions are used, the autosomal ("female") homologues of the Y repeat should be readily detectable as well as the characteristic spot associated with the Y chromosome (Burns *et al.*, 1988); failure to do so usually indicates that more vigorous proteolytic digestion is required. While it could be argued that the degree of digestion necessary to reveal chromosomal sequences is probably greater than that required to detect "free" nuclear or cytoplasmic viral nucleic acid, such a control is particularly relevant when probing for integrated viruses, and we feel that this is still the most reliable method available of ensuring that reasonable confidence can be attached to a negative result. An alternative method for judging the sensitivity of a protocol involves the use of cell lines with known numbers of integrated viral sequences as standards (i.e. HeLa cells containing 10–50 copies of HPV18, or Raji cells containing 50–100 copies of Epstein-Barr virus). Such cells can only serve as a valid comparison, however, if they have been fixed and processed in an identical manner to the tissue under investigation. A further method of assessing the sensitivity of a protocol involves the use of routinely processed tissues from transgenic animals whose cells contain a known number of copies of a particular sequence. We have employed transgenic mice containing between 2 and 20 copies of the human alpha-1-antitrypsin gene per cell to assess the sensitivity of non-isotopic *in situ* hybridization using digoxigenin labelled DNA probes

(Fleming *et al.*, 1991). As few as 10 copies of a 1.3 kb sequence could be unequivocally detected, suggesting that the non-isotopic detection of low-copy integrated viral sequences in formalin-fixed paraffin-embedded tissues should be possible.

1.1.4 Stringency conditions

Viruses which are related but not identical to the probe can be detected by altering the stringency conditions of the reaction (see Chapter 1, Bornkamm *et al.* (1983) and Myerson (1988) for discussions of stringency calculation). Under conditions of low stringency, nucleic acid sequences with a degree of mismatch can hybridize. This finding has been employed in the detection of new viral strains and classification of related viruses, however sequence homologies also exist between parts of the human genome and numerous viruses, which under inappropriate stringency conditions can lead to false positive results. Conversely, the use of excessively high stringency conditions will lead to loss of sensitivity. It is thus important that stringency conditions be carefully calculated and applied in order to suit the particular situation under investigation.

1.1.5 Further important methodological considerations

Further important methodological considerations relate to problems of artefactual probe binding and non-specific background signal. These can result from a variety of factors both intrinsic to the hybridization protocol, and intrinsic to the tissue under study (see Myerson, 1988). The most intransigent cause of background in our experience of non-isotopic *in situ* hybridization using biotinylated probes is the presence of endogenous biotin in "hypermetabolic" tissues such as liver, kidney and pancreas. This endogenous biotin is bound by avidin or anti-biotin antibodies employed in probe detection, and can lead to highly deceptive false positive signals. Although some authors report success in blocking or removing endogenous biotin (Grody *et al.*, 1987; Naoumov *et al.*, 1988a) we have found that none of these methods enable reliable and reproducible detection of low copy viral genomes in these tissues. For that reason this laboratory is currently concentrating on the development of protocols employing alternative non-isotopic labels such as digoxigenin (Fleming *et al.*, 1991; Morey *et al.*, 1991b; Figure 2).

1.1.6 Controls

The use of appropriate positive and negative controls is crucial to the success and reliability of *in situ* hybridization as a diagnostic or investigative tool. If it is to assume even a minor role in routine laboratory diagnosis of viral infections (where decisions on patient management will depend on the result), it is imperative that reliable controls are employed. Positive controls should in-

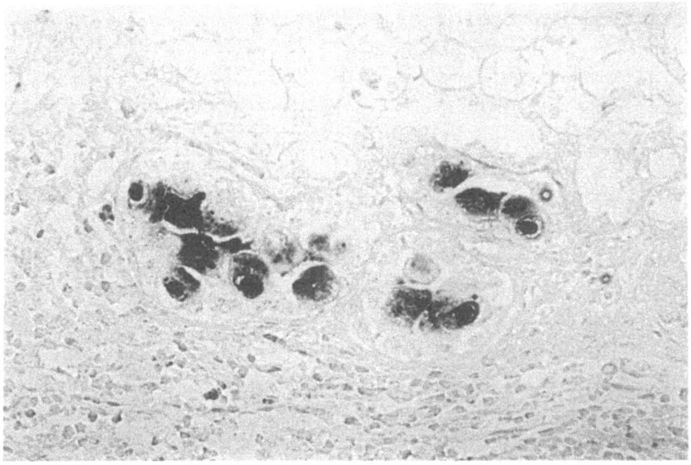

Figure 2 Cytomegalovirus DNA localised within the renal tubules in a case of fetal CMV infection. The DNA probe was nick-translated with digoxigenin and detected with an anti-digoxigenin alkaline phosphatase conjugate and NBT-BCIP chromogenic substrate. CMV DNA is seen in both nuclei and cytoplasm of cytomegalic epithelial cells.

clude tissues known to be positive for the virus using other means (but processed and tested in the same manner as the tissue under investigation) and general marker probes which can control for the efficiency of the hybridization reaction on each occasion (such as probes for the Y repeat or other high-copy genomic sequences). Negative controls include the use of labelled vector sequences without the viral insert, non-related similarly-labelled probes, opposite sense strand probes (if riboprobes are used) and the use of tissues reliably known to be negative for the virus. DNAse and RNAse digestions should also be performed in order to verify the identity of the target nucleic acid species being detected. In cases where there is any doubt as to the specificity of the reaction it is necessary to confirm the identity of the viral nucleic acid using dot-blot hybridization or other methods such as the polymerase chain reaction (Salimans *et al.*, 1989; Pezella *et al.*, 1989).

2. DIAGNOSTIC APPLICATIONS

The use of *in situ* hybridization in viral diagnosis is especially relevant for situations in which the virus cannot be easily or rapidly detected by other means. This includes viruses which grow insufficiently or not at all in cell culture (possibly requiring specific host cell factors which are absent in standard cell lines), and those which escape detection by immunological methods. This may be due to protein products which are antigenically variable, or synthesized at levels below the threshold for detection by serology or immunohistology. The latter situation frequently arises in clinically inapparent "latent" and "slow" infections. Even if antigen production does occur, and appropriate antibodies are available, in cases where only fixed tissue is available immunohistological studies may be unsuccessful due to masking or destruction of the antigenic determinants. Nucleic acids appear to be largely unaffected by fixation however, and are readily rendered accessible for study by proteolytic digestion. *In situ* hybridization can be an effective diagnostic tool where rapid diagnosis is required, or where the state of the tissue precludes other forms of investigation (for example because of autolysis, or contamination by bacteria or other viruses). It also can be used to quantitate absolute numbers of infected cells, and to monitor the effects of antiviral treatment at the tissue level. Low stringency hybridization using broadly reacting probes from conserved regions of viral genomes can be used to detect related viruses which may be overlooked by immunohistochemical methods. A diagnostic application which is unique to *in situ* hybridization is the ability to identify viral subtypes (as with the various types of human papillomavirus) in relationship to associated histopathological abnormalities. The technique has also been applied to retrospective searches for viral involvement in a variety of diseases of unknown aetiology.

2.1 Sensitivity

An important feature of *in situ* hybridization is its sensitivity in instances where only a small proportion of cells are infected; this may occur in early, focal, latent or resolving infections. Examination of such cases using tissue extraction methods such as dot hybridization is likely to be negative (due to "dilution" of the target viral sequences by cellular nucleic acid), electron microscopic searches for intact viral particles have a limited chance of success, and the extent of antigen expression may be sufficiently reduced as to be below the threshold for detection. In such cases, *in situ* hybridization (which can theoretically detect the presence of specific hybridization signal in a single cell per section) may enable a diagnosis to be made. An example of such sensitivity was provided by Brahic *et al.* (1981) in an early study on visna (a slow virus which causes progressive neurological deterioration in sheep); in choroid plexus cells from experimentally infected sheep, viral antigen was almost undetectable, however by *in situ* hybridization with radiolabelled probe, visna RNA and proviral DNA were detected in between 1–3% of cells. While the polymerase chain reaction is generally believed to be an even more sensitive method of viral detection in such cases of low-level infection, standard protocols achieve this at the expense of histological localization of the virus, and the presence of "inhibitors" in paraffin-embedded material can limit the sensitivity of the technique in routinely processed specimens. Successful combination of the amplifying power of PCR with the localizing ability of *in situ* hybridization has only recently been reported (Haase *et al.*, 1990). The antithetical demands of PCR (for good access to target) and *in situ* hybridization (for retention of nucleic acid in cellular structures) were satisfied by the use of multiple primer pairs to amplify overlapping segments of DNA. The cohesive termini of the amplified fragments enabled the formation of large covalently-linked segments which were effectively retained within the cell. Applied to the detection of visna virus DNA in experimentally infected cultured cells, the "*in situ* amplification" technique led to 200-fold improvement in sensitivity (Haase *et al.*, 1990). While the technique has not, as yet, been successfully applied to formalin-fixed, paraffin embedded tissues, it obviously holds considerable promise for use in investigations into the pathogenesis of viral infections. In particular, the amplification of target within single cells is inherently less likely to be affected by "contamination" by exogenous nucleic acids, a major problem with the use of the polymerase chain reaction in solution.

2.2 Rapidity

Because of the relative rapidity of modern *in situ* hybridization techniques compared to traditional viral culture, *in situ* hybridization (especially using non-isotopic probes) has been advocated as a diagnostic tool in certain cases where patient management is dependent on rapid diagnosis. This is particularly pertinent now that anti-viral chemotherapy is becoming a clinical reality. The diagnosis of herpes simplex virus (HSV) encephalitis using cells obtained from the cerebrospinal fluid (Bamborschke *et al.*, 1990) is an example of such an application. Viral DNA was detectable in cytospun preparations of these cells at the time of onset of symptoms and before the beginning of intrathecal IgG synthesis. A second example involves the use of *in situ* hybridization to determine whether cytomegalovirus (CMV) infection is responsible for hepatic dysfunction after liver transplantation. Infection with CMV frequently complicates the clinical course of patients on immunosuppressive treatment, and may be directly responsible for graft failure, however it is vitally important to differentiate infection from rejection of a transplanted organ. Immunosuppressed patients may lack the usual serological markers of infection, or have co-existent but clinically insignificant reactivation of previous infection. Direct demonstration of CMV DNA in areas of inflammation, however, provides strong evidence that a case of post-transplant hepatitis is due to the virus

(Naoumov *et al.*, 1988b; Masih *et al.*, 1988). A diagnosis can be obtained in less than 24 hours (as opposed to viral culture which may take weeks to show a cytopathic effect), thus providing the opportunity for early adjustment of management. The recent availability of specific oligonucleotide probes for the detection of mRNA from the abundantly transcribed Immediate Early and Early CMV genes (British Biotechnology Ltd.) holds promise for rapid diagnosis at even earlier stages of infection.

2.3 Diagnosis of non-culturable viruses

In situ hybridization has proved a particularly useful tool in the diagnosis and assessment of infections caused by viruses with stringent host cell requirements which do not permit the virus to replicate effectively in continuous cell-line culture systems. Two such viruses which have been studied in this laboratory are the human papilloma virus (HPV) and the human parvovirus (B19).

The human papilloma virus (HPV) propagates effectively only in terminally differentiated keratinocytes. Standard cell lines (which by nature are relatively undifferentiated) cannot support the growth of this virus. Prior to the advent of recombinant DNA technology HPV variants were suspected on clinical grounds, however cloning and sequencing techniques have now enabled the identification of over 60 different subtypes. Antibody to the highly conserved capsid protein allows immunohistological detection of productively infected cells, but examination of viral nucleic acid remains the only effective method of diagnosing latent infection. In infected epithelia, the accumulation of viral DNA and the expression of antigen both correlate with the degree of keratinization, though cells positive for HPV DNA exceed the number containing capsid antigen and are found at lower levels of the epithelium (Beckmann *et al.*, 1985). Though efforts are underway to produce type-specific monoclonal antibodies (Patel *et al.*, 1989), DNA hybridization using type-specific nucleic acid probes remains the only reliable means of differentiating the vast majority of HPV types. If low stringency conditions are employed in the hybridization reaction, cross hybridization between various types can occur. Such cross hybridization is proportional to the degree of similarity in the base pair arrangement of the target nucleic acid and the probe, and is frequently seen between types 6 and 11 which share a high degree of sequence homology. If hybridization and washing are performed under conditions of sufficiently high stringency however, positive identification of the particular type present in a given tissue is usually possible (Herrington *et al.*, 1990). This is of some importance as a large body of epidemiological evidence suggests that certain papillomavirus types are associated with infection at particular anatomical sites, and with the development of various epithelial dysplasias and neoplasias (reviewed by Syrjänen, 1987). Common skin warts are usually due to infection with HPV types 1 to 4, while types 6, 11, 16 and 18 are found in anogenital lesions and are presumed to be sexually transmitted. Types 6 and 11 have been generally associated with benign condylomas and low grade dysplasias, but types 16 and 18 (as well as the less common types 31, 33 and 35) have been more frequently found in cases of severe genital dysplasia and cervical carcinoma. While these associations are far from absolute, and an aetiological role for the virus in the production of malignancy has not been proven, a large number of researchers have used *in situ* hybridization techniques to confirm the findings obtained by other methods and to investigate further the role of various HPV subtypes in a wide variety of epithelial lesions (e.g. Beckmann *et al.*, 1985; Burns *et al.*, 1987; Beckmann *et al.*, 1989; Terry *et al.*, 1989).

High stringency *in situ* hybridization has been employed in this laboratory in a study of HPV types in anogenital warts in children (Padel *et al.*, 1990). As genital HPV infection is a sexually transmitted disease in adults, it has been suggested that the presence of genital warts in a child is sufficient grounds for investigating that child as a possible victim of sexual abuse. It has been previously reported that genital warts in children are almost exclusively associated with the adult "genital" types (6,11,16 and18)(Vallejos *et al.*, 1987). Results from this laboratory do not support that finding. Using type-specific double stranded DNA probes labelled with biotin and an avidin-alkaline phosphatase detection system, it was found that genital warts from 6 out of 17 children examined contained "skin" HPV types (2 or 3). In two cases, warts containing virus of the same type were present on the child's hands, raising the possibility of self-inoculation as the route of viral transmission. The appearance and site of the anogenital warts did not correlate with HPV type. These findings have important implications, especially on consideration of the problems associated with an incorrect diagnosis of sexual abuse.

The human parvovirus (B19) has recently been shown to be an important and widespread human pathogen (reviewed by Goldfarb, 1989). The causative agent of erythema infectiosum, the virus has also been shown to be the major cause of aplastic crises in patients with chronic haemolytic anaemia. It is capable of causing chronic bone marrow dysfunction in the immunosuppressed, and infection during pregnancy can lead to fetal hydrops and stillbirth. Because of requirements for as yet unspecified host cell factors, the human parvovirus shows a high degree of target cell specificity, with a particular predilection for red blood cell precursors. Though the virus has been propagated in short term haemopoietic cell suspension culture, it has not been successfully grown in continuous culture.

We have used *in situ* hybridization with non-isotopic probes to diagnose

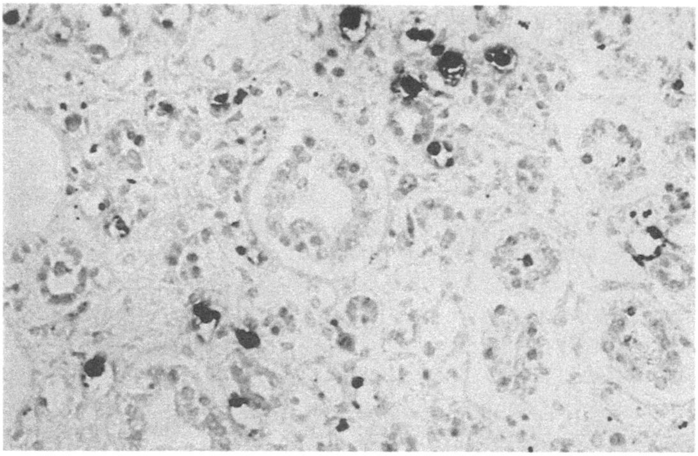

Figure 3 Parvovirus B19 infected cells in small vessels in the kidney of a fetus with non-immune hydrops. Digoxigenin labelled DNA probe detected with anti-digoxigenin alkaline phosphatase conjugate and NBT-BCIP chromogenic substrate.

human parvovirus infection in archival material from stillborn fetuses (Porter *et al.*, 1988; Morey *et al.*, 1991b). Tissue from such cases is frequently difficult to examine using standard techniques because of maceration and autolytic damage, however viral DNA is relatively resistant to degradation and can be readily demonstrated in infected cells within fetal tissues (Figure 3). Parvovirus DNA was detected in routinely processed tissues from 10 out of 37 cases of otherwise unexplained non-immune fetal hydrops, accounting for 8% of all cases of non-immune hydrops examined over a 16 year period. Cells containing parvoviral DNA were distributed within blood vessels and alveoli, the number of positive cells exceeding the number of cells found to contain typical viral inclusions on routine light microscopy. The question of target cell specificity of human parvovirus in these cases has also been examined by a combination of immunohistology for cellular antigens and *in situ* hybridization for parvoviral DNA (see below).

2.4 Diagnosis of latent or slow virus infections

In situ hybridization remains the only effective method of diagnosing latent infections in a histological context, as viral protein production is often not only insufficient to stimulate an effective immune response, but also below the threshold for detection by immunological techniques. An example of the application of *in situ* hybridization techniques to the investigation of latent viral disease which has important and disturbing implications for public health planning includes the detection of human immunodeficiency virus (HIV-I) nucleic acid in a small minority of peripheral blood mononuclear cells of patients who were "at risk" of infection but seronegative at the time of the study (Pezella *et al.*, 1989). Additional examples include the localization of varicella zoster virus (VZ) and herpes simplex virus (HSV) transcripts in the neural ganglia of patients without evidence of overt disease (Croen *et al.*, 1988), and the detection of measles virus genomes in brain biopsies of patients with subacute sclerosing panencephalitis (SSPE) (Haase *et al.*, 1985b). *In situ* hybridization has also been used in studying animal models of latent and slow viral disease (e.g. Brahic *et al.*, 1981, 1984; Peluso *et al.*, 1985; Stowring *et al.*, 1985; Gendelman *et al.*, 1986), providing a powerful means of investigating the mechanisms behind the establishment and maintenance of latent infection (see below).

2.5 Search for viral involvement in cryptogenic disease

The realization that viral infection can occur (and be detected by *in situ* hybridization) in the absence of any of the conventional markers of viral disease, has encouraged numerous groups to use the technique to search retrospectively for viral involvement in a range of diseases of unknown aetiology. Many of these investigations have produced a negative result; they include studies which have failed to demonstrate a significant relationship between adenovirus infection and follicular bronchiectasis (Hogg *et al.*, 1989), or between CMV and Kaposi's sarcoma (Grody *et al.*, 1988). Other studies have detected specific viral nucleic acid in an equivalent or greater number of "control" than "affected" subjects. They include investigations into the role of Epstein Barr virus (EBV) in Sjögren's syndrome (Venables *et al.*, 1989) and of measles virus in multiple sclerosis (Haase *et al.*, 1984). Such findings obviously argue against a simple causal relationship between the particular viral infection and the disease in question, and demonstrate that humans may normally be host to a variety of ubiquitous viral genomes without stimulation of an immune response or apparent evidence of disease. The possibility remains that pathological changes are brought about by an atypical host response to the presence of a common infection.

Of particular interest however, are several studies which have suggested a significant association between the presence of viral genomes and the occurrence of disease. They include studies which have correlated the presence of HSV and CMV residues with atheromatous change in arterial walls (Benditt *et al.*, 1983; Yamashiroya *et al.*, 1988) – a finding given additional significance by the known association of Marek's disease virus (a herpesvirus) and the production of atheroma in birds – and the demonstration of CMV nucleic acid in pancreatic tissue from a subset of patients with non-insulin dependent diabetes mellitus (Lohr *et al.*, 1990). In the latter study, the hybridization signal was localized primarily to the islets of Langerhans, but was not associated with any morphological abnormality or evidence of inflammation. Such preliminary findings obviously require further confirmation and the precise significance of the presence of these viral sequences remains to be determined, however *in situ* hybridization would seem to be the technique best suited to explore such questions.

3. *IN SITU* HYBRIDIZATION IN THE STUDY OF VIRUS-HOST CELL INTERACTIONS

In situ hybridization provides a versatile method of investigating virus-host interactions at every level; from whole animal studies of organ involvement with a particular virus, to analysis of sites of viral integration into host cell chromosomes. The technique has provided considerable insight into the pathogenesis of disease due to certain viral infections by allowing positive identification and characterization of the cell types involved. This has led to a re-evaluation of the mechanisms of spread of certain viruses through the body and to a better understanding of the role of the immune system (as opposed to direct viral cytopathic effect) in the resultant tissue damage. *In situ* hybridization studies have provided some unique insights into the process of latent infection and are assuming a significant role in the study of virus-related tumours.

3.1 Identification of host cells and mechanisms of viral spread

A variety of ingenious techniques have been evolved to enable the study of viruses at the whole animal or organ level. Most have involved cutting thick sections onto some form of adherent tape, then detecting viral nucleic acid using high energy radiolabelled probes (see review by Lipkin *et al.*, 1990). For such applications, detailed resolution is not required, and a very satisfactory picture of the pattern of viral involvement at the macroscopic level can be obtained. Such macroscopic hybridization techniques can be useful to determine whether infection is diffuse or localized, to follow the anatomical distribution of a virus over the course of an infection, and to guide the choice of material for microscopic studies. Examples of the application of this concept include detection of visna virus in the paraventricular areas of the sheep brain (Haase *et al.*, 1985c), localization of ground squirrel hepatitis virus to the liver in the chipmunk (Lipkin *et al.*, 1990), and visualization of yellow fever virus in the abdominal fat body of infected mosquitoes (Ballinger *et al.*, 1988).

It is at the tissue level, however, that *in situ* hybridization has found its widest application. The technique has been instrumental in identifying the host cell type and mode of spread in a wide range of viral infections. For example, peripheral blood leukocytes have been found to play a role in the spread of hepatitis B virus (HBV) (Hadchouel *et al.*, 1988), CMV (Turtinen *et al.*, 1987) and HIV-I (Pezzella *et al.*, 1989). The affected cells in progressive multifocal leucoencephalopathy (PMLE), a demyelinating condition associated with reactivation of latent JC papovavirus in immunosuppressed patients, have been shown to be oligodendrocytes and astrocytes (Aksamit *et al.*, 1985). In fetal and placental tissue removed from HIV-positive pregnant women at 8 weeks gestation, HIV-I nucleic acid has been detected in maternal decidual lymphocytes, villous trophoblast and Hofbauer cells, villous capillary endothelium and embryonic blood cell precursors (Lewis *et al.*, 1990), suggesting a pathway for cell-to-cell transmission of the virus to the fetus and confirming that fetal infection can occur early in pregnancy. Demonstration of human herpes virus 6 (HHV-6) in submandibular salivary gland tissue (Fox *et al.*, 1990) supports the hypothesis that this virus may persist in the salivary glands and be transmitted by saliva. *In situ* hybridization has also been used extensively to study mechanisms of viral spread in animal models. For example, transynaptic neuronal spread has been shown be involved in the entry of murine hepatitis virus into the CNS of experimentally infected mice (Perlman *et al.*, 1989) and evidence for venereal transmission of murine CMV was provided by the detection of viral nucleic acid in spermatocytes (Baskar *et al.*, 1986). Replication of the duck hepatitis B virus has been found to occur in pancreatic islet cells as well as in hepatocytes (Jilbert *et al.*, 1987). Strand-specific probes have proved very useful in studies of Aleutian mink disease parvovirus infection (Alexandersen *et al.*, 1987); because this parvovirus encapsidates single-stranded DNA of only one polarity as the virion genome, a probe of opposite polarity to the genome will detect both virions and double-stranded replicative intermediates, while a probe of the same polarity as the genome will hybridize only to replicative form DNA. The use of such probes has enabled the differentiation of cells supporting viral replication from cells that have merely sequestered the virus, and has shown that alveolar lining cells are the major site of viral replication in infected newborn mink (Alexandersen *et al.*, 1987).

Studies on several viruses have indicated that the number of cells containing viral nucleic acid by *in situ* hybridization exceeds the number of cells showing morphological evidence of infection, indicating that appearances under light microscopy after routine staining can give a misleading impression of the extent of infection. This has been a consistent finding in studies of CMV infection in many different organs (Myerson *et al.*, 1984; McDougall *et al.*, 1986; Naoumov *et al.*, 1988b; Porter *et al.*, 1990), and a similar pattern of infection with human parvovirus B19 was seen in this laboratory (Porter *et al.*, 1988). In cases where tissue preservation is not ideal or a heterogeneous cell population is present, *in situ* hybridization for viral nucleic acid can be combined with immunohistology for cellular antigens to allow unequivocal identification of infected cell types. This combination was first reported in a study on the slow virus visna in sheep (Stowring *et al.*, 1985). Using a ^{3}H labelled probe for visna RNA and cell-type specific antisera detected by the immunoperoxidase method, infected cells in the brain were identified as oligodendrocytes. A similar technique was used to demonstrate that visna virus can infect mononuclear-phagocytic cells, the extent of viral replication being dependent on the degree of differentiation toward the macrophage phenotype (Gendelman *et al.*, 1986). It was also employed to demonstrate that HIV-I was mainly found within macrophages in the central nervous system of patients with the acquired immunodeficiency syndrome (Koening *et al.*, 1986; Eilbott *et al.*, 1989). In a variation on this protocol, Vaseux *et al.* (1990) have combined immunolabelling for HIV-I antigen with *in situ* hybridization for JC papovavirus in brain tissue from AIDS patients with PMLE and severe HIV-I encephalitis. They were able to show that different cell populations were affected by the two viruses, but postulated that the encephalitis had spread via the recruitment of HIV-I infected macrophages into areas of JC-induced demyelination. Non-isotopic variants of the double-labelling technique using biotinylated probes and immunoperoxidase detection of antigens have recently been used in frozen sections or cultured cells to detect HPV in cytokeratin-positive cells and CMV in vimentin-positive cells (Mullink *et al.*, 1989b; van der Loos *et al.*, 1989). Most authors have advised performing immunolabelling first, though if the antigen is heat stable it may in some instances be preferable to perform the *in situ* hybridization first (Brahic and Haase, 1989).

Combined non-isotopic protocols suitable for use on paraffin-embedded material employing immunohistology for specific cellular antigens using the alkaline phosphatase–anti-alkaline phosphatase (APAAP) method (Cordell *et al.*, 1984) and biotin- or digoxigenin-labelled probes for viral nucleic acid have been developed in this laboratory (Porter *et al.*, 1990; Morey *et al.*, 1991b). The APAAP method is generally recognized to be more sensitive than the immunoperoxidase labelling, and can be used prior to routine *in situ* hybridization without alteration to either technique or any additional steps between them. The use of this double-labelling technique has enabled positive identification of infected host cells in cases of human fetal parvovirus infection (Porter *et al.*, 1990; Morey *et al.*, 1991b). This had been difficult with standard counterstains because of the poor state of preservation of the tissues. Although tissue autolysis can adversely affect the expression of various cell surface and cytoskeletal antigens, judicious selection of antibodies can minimize these effects. We found that while the majority of cells containing parvovirus DNA were of erythroid lineage, hybridization signal was also found within numerous cells labelled as mononuclear-phagocytes (Colour plate 5), and within some myocardial cells labelled with an antibody to desmin. This suggests that the heart failure and hydrops in infected fetuses may not be due to anaemia alone, but may also reflect direct damage to the myocardium.

We have also employed combined immunohistology and *in situ* hybridization using digoxigenin labelled probes to investigate the host cell specificity of parvovirus in a short term haemopoietic cell suspension culture system (Morey *et al.*, 1991c). Evidence of viral replication was found only in cells labelling with erythroid markers, and the pattern of distribution of the virus was dependent on the stage of maturation of the cell. The disruptive effect of infection on the host cell chromatin was demonstrated by combining double *in situ* hybridization (a digoxigenin labelled probe for whole human DNA together with a biotin labelled probe for parvovirus visualized using different enzymatic substrates) with immunolabelling of cell type specific antigens, giving triple colour detection of three specific targets in a single cell (Morey *et al.*, 1991a: Colour plate 4)

3.2 Viral gene expression and mechanisms of disease

While numerous studies have compared sensitivity of immunohistological detection of viral antigen with labelling of viral nucleic acid using *in situ* hybridization, a more useful technique involves combining the protocols to enable the detection of viral genes or mRNA and their products in single cells. The relationship between the presence of viral nucleic acid and the expression of viral antigen has been investigated using combined protocols for viruses including HBV (Blum *et al.*, 1984), Theilers murine encephalitis virus (Brahic *et al.*, 1984), visna (Gendelman *et al.*, 1985), measles (Gendelman *et al.*, 1985), HPV (Mullink *et al.*, 1989b; Cubie and Norval, 1989), HSV (Kennedy *et al.*, 1988), hepatitis delta virus (Negro *et al.*, 1989) and CMV (Wolber and Lloyd, 1988; van der Loos *et al.*, 1989; Porter *et al.*, 1990). While one group has reported that CMV antigen detection is unaffected by a preceding hybridization procedure (Wolber and Lloyd, 1988), most groups have concluded that it is prudent to perform immunohistological staining prior to *in situ* hybridization, in case denaturing temperatures adversely affect the antigen. A consistent finding in many of these combined studies is the demonstration of viral genomes in a greater proportion of cells than contain viral antigen. Such results can provide important insights into the relationship between viral infection, antigen expression and the presence of pathological changes. For example, the absence of protein may identify a point of blockade in the viral life cycle, while an abundance of protein products in a cell has important implications for the stimulation of an immune response.

Combined labelling procedures have been particularly useful in studies of latent viral infections. Double labelling for viral genes and antigens has been used extensively by Haase and co-workers to support the thesis that "persistent" infections are associated with restricted viral gene expression which enables the virus to escape detection by the immune system for prolonged periods (Haase, 1986; Brahic and Haase, 1989 for reviews). A low level of antigen production was also postulated to be important for the spread of visna virus through the body of experimentally infected sheep; low copy visna genomes were found in monocytes in the cerebrospinal fluid, despite the presence of neutralizing antibody (Peluso *et al.*, 1985). It was suggested that the low level of viral gene expression in these cells enabled them to evade immune attack while disseminating the virus (the "Trojan Horse" theory). A similar mechanism was proposed for the persistence of HBV in cases of chronic active hepatitis; most hepatocytes found to contain HBV DNA did not express hepatitis B core antigen (Blum *et al.*, 1984), thus an immune response directed against the antigen would be ineffective in eradicating the virus. Interestingly, in both chronic HBV infection (Blum *et al.*, 1984) and chronic hepatitis delta infection (Negro *et al.*, 1989), cells undergoing degenerative change did not appear to contain viral antigen and/or viral nucleic acid, though infected cells were found adjacent to foci of inflammation. This supports the possibility that the viral infection may have initiated an autoimmune process which was responsible for the observed tissue damage.

As a virus may cause disturbance by inducing or repressing the transcription of certain host genes, the ability to directly assess the effects of viral infection on host cell nucleic acid metabolism would of obvious benefit. Using a biotin labelled probe for Theiler's disease murine picornavirus mRNA and a ^{35}S labelled probe for β-actin mRNA, Ozden *et al.* (1990) have recently co-detected viral and cellular mRNAs in single infected cultured cells. While in that study no obvious difference was seen in the expression of the b-actin gene between infected and non-infected cells, the technique has considerable potential for investigating the effects of infection on cellular gene transcription, especially when only a subpopulation of cells are affected.

The mechanisms behind the maintenance of latent infection of neural tissue by herpes simplex virus (HSV) have been extensively investigated using strand-specific probes for particular RNA transcripts, the amount of latent DNA itself being insufficient for detection by current *in situ* hybridization protocols. Studies on neural ganglia from a number of species including humans have demonstrated that latent infection is confined to neurons. It is associated with the expression of a single family of "latency associated transcripts" which map to the region of the viral genome coding for the immediate early protein (ICPO), but which unexpectedly are complementary to ICPO mRNA, suggesting a possible inhibitory role (see review by Stevens, 1989). The precise role of these novel transcripts in the maintenance of, or reactivation from, the latent state is still under investigation. By contrast, latent infection with varicella zoster virus (VZV) has been shown to involve non-neuronal cells, with multiple regions of the genome being expressed (Croen *et al.*, 1988). This finding may underlie the different clinical features of disease produced by these viruses following reactivation. Further application of techniques for investigating the expression of particular genes at the cellular level will undoubtedly provide insight into the mechanisms associated with latent infection by many other viruses.

3.3 Viral involvement in tumourigenesis

Members of several viral families have been epidemiologically associated with the development of tumours. While simple causal associations are difficult to make, *in situ* hybridization has proved a very useful tool for investigating the relationship between viral infection and neoplastic change at the tissue level. For example, the technique has been used to identify various types of papillomavirus in a wide range of malignant and pre-malignant epithelial lesions in humans and other species (Milde and Loning, 1986; Syrjänen *et al.*, 1988; Beckman *et al.*, 1989; reviews by Syrjänen, 1987; Crum and Roche, 1990). The ability to detect low levels of viral DNA in dysplastic, less differentiated cells which do not express viral antigen is a major advantage of the technique. HPV has been detected not only in primary lesions, but also in metastatic deposits of cervical carcinoma in distant lymph nodes (Lewandowski *et al.*, 1990). Epstein Barr virus genomes have been detected in malignant cells in biopsies of nasopharyngeal carcinoma (Hawkins *et al.*, 1990) and several types of lymphoproliferative lesions including Hodgkin's and non-Hodgkin's lymphoma (Hamilton-Dutoit *et al.*, 1989; Weiss *et al.*, 1989), T cell lymphoma (Jones *et al.*, 1988) and hairy cell leukaemia (Wolf *et al.*, 1990), the association of EBV with lymphoproliferative lesions being especially common in immunosuppressed patients. The detection of hepatitis B virus DNA in both surface antigen-positive and surface antigen-negative patients with hepatocellular carcinoma has been reported (Brambilla *et al.*, 1986).

A major advantage of using *in situ* hybridization in such studies is that it enables correlation of viral presence with the degree of cellular dysplasia, and permits investigation of abnormalities of cellular function consequent upon infection. For example, *in situ* hybridization has been combined with immunohistological studies to determine the relationship between viral infection and the loss of normal cellular differentiation characteristics. Differences in the pattern of cytokeratin expression between dysplastic and non-dysplastic cervical lesions associated with HPV infection have been postulated to reflect transformation of the cells by the virus (Syrjänen *et al.*, 1988). Protocols enabling the differential detection of various HPV RNA transcripts in precancerous lesions (Crum *et al.*, 1988) may provide significant insights into the role of certain transcripts in initiation of neoplastic change.

Some early *in situ* studies suggested that herpes simplex virus (HSV) RNA may be present in a considerable proportion of cervical cancers and precancers (McDougall *et al.*, 1980; Maitland *et al.*, 1981). While recent attention has been focused on the role of human papillomavirus in such lesions, the occasional detection of both HSV and HPV nucleic acid in the same tumour raises the possibility that these viruses may have an initiator-promoter relationship in the production of some tumours (McDougall *et al.*, 1986). This area has still to be fully explored, however simplified non-isotopic techniques for the simultaneous detection of two viral genomes within single cells (Mullink *et al.*, 1989a) theoretically offer an ideal way of investigating such questions.

3.4 Chromosomal integration of viral DNA

A number of viruses have the ability to integrate copies of their genomes into host cell chromosomes. While a normal event in the life-cycle of retroviruses such as HIV-I, this can also occur with several non-retroviruses including HPV, EBV and HBV. Integration not only makes possible the establishment of latent infection, but also appears to be implicated in the transformation of host cells to an immortal or malignant phenotype. Immortalized cell lines containing integrated copies of all the above viruses are available, and metaphase spreads of chromosomes from such cells have been investigated using *in situ* hybridization to determine the chromosomal sites of viral integration and their relationship if any, to known oncogenes or "fragile sites" (Henderson *et al.*, 1983; Popescu and Di Paolo, 1990). The majority of such studies have employed ^{3}H labelled probes, however improvements in non-isotopic protocols have recently enabled more rapid detection of viral integration sites with improved resolution (Teo and Griffin, 1987; Lawrence *et al.*, 1988; Lawrence *et al.*, 1990). For example, using fluorescent detection of biotin-labelled probes, Lawrence *et al.* (1988) have visualized two integrated copies of EBV on chromosome 1 in metaphase spreads of the Namalwa cell line. Using probes to the different ends of the genome they compared the location of the integrated sequences in metaphase and interphase cells and were able to determine that the two copies lay in opposite orientation and were separated by about 340 kb of chromosomal DNA. Such sensitive labelling strategies combined with new techniques for chromosome karyotyping using non-isotopic probes for interspersed repetitive sequences (Boyle *et al.*, 1990) offer the exciting possibility of viral gene localization on simultaneously-banded chromosomes.

4. STUDIES OF VIRAL LIFE CYCLES WITHIN CELLS

A final area of application for *in situ* hybridization techniques is as an aid to understanding the basic processes of infection and viral replication within individual cells. For example using fluorescent non-isotopic techniques, Lawrence *et al.* (1990) have detected a single bright nuclear focus of newly transcribed HIV-I RNA only 12 hours after experimental infection; its appearance preceded the accumulation of nucleic acid throughout the rest of the cell and was similar to that found in lymphocytes from HIV infected patients and in a cell line with a single integrated copy of defective HIV provirus. This suggests that the nuclear focus was derived from single or few copy viral genomes per cell, and that such an *in vitro* system may be useful for assessing the effects of various treatment strategies on the replicative activity of the virus (Lawrence *et al.*, 1990). Using similar techniques, Lawrence *et al.* (1989) have reported the restriction of EBV RNA transcripts to a "track" extending from the interior to the periphery of infected nuclei, thus providing evidence for the non-random organization of genes and their transcripts in interphase nuclei.

The advent of the confocal laser scanning microscope for optically sectioning cells has given a major impetus to the study of intranuclear topology. The ability to generate 3-D reconstructions of the spatial arrangement of viral genomes within nuclei has been employed in a study of HeLa cells infected with adenovirus-2 and minute virus of mouse (MVM) parvovirus (Moen *et al.*,1990). Using a non-isotopic fluorescent labelling system, MVM replication was localized to the nucleoli, while adenovirus-2 DNA was found in multiple non-nucleolar "replication factories" anchored along the nuclear envelope. In cells co-infected with the two viruses, MVM DNA replication was more efficient, and found to be recompartmentalized to the adenovirus replication factories. Confocal microscopy and 3-D reconstruction was also used by Harders *et al.* (1990) to demonstrate that potato spindle tuber viroids and their replicative intermediates were homogeneously distributed throughout the nucleoli in infected tomato leaf cell nuclei, a feature which could not have been determined by standard microscopy.

4.1 Ultrastructural studies

Relatively little work has been done on the intracellular localization of viral nucleic acid at the electron microscope level. Early attempts at ultrastructural *in situ* hybridization with radiolabelled probes were hampered by poor resolution and long processing times. It is only quite recently with the advent of suitable electron dense, non-radioactive labels that viral genomes have been effectively localized at the ultrastructural level. Wolber *et al.* (1988, 1989) have employed a pre-embedding *in situ* hybridization technique to study the pathogenesis of HSV and CMV infection in cultured fibroblasts using biotinylated probes detected with streptavidin conjugated to colloidal gold spheres. CMV DNA was detected at the edge of electron dense regions in viral inclusions in fibroblast nuclei, while HSV RNA was demonstrated adjacent to the nuclear envelope. In neither case were gold clusters found bound to viral nucleocapsids, possibly because of difficulties with penetration of the gold label. Puvion-Dutilleul and co-workers have very successfully utilized an alternative strategy, performing *in situ* hybridization on thin sections of HSV infected cells embedded in Lowicryl resin (Puvion-Dutilleul and Puvion, 1989a; Puvion-Dutilleul *et al.*, 1989). Using a biotinylated probe visualized by immunogold labelling they were able to follow the early events of infection and to demonstrate that while non-encapsidated (i.e. replicating) HSV DNA was found exclusively in the virus-induced central compartment of the nucleus, viral DNA elsewhere in the cell was always associated with capsids. Although these are preliminary findings, ultrastructural localization of viral nucleic acid represents one of the most exciting applications of the *in situ* hybridization technique currently being explored, with potential for investigating the processes involved in viral infection and replication at the most fundamental level.

6. CONCLUSIONS

The use of *in situ* hybridization techniques in the study of viral disease has undoubtedly been one of the most fruitful applications of the technique. It has a major role to play in the localization of viruses which evade detection by other means and has provided important data on the mechanisms of spread of certain viruses, and the processes involved in tissue damage secondary to infection. It also has considerable potential for investigating the basic biology of viral infection and replication. Several recent technological advances are likely to have a major impact on future applications of the technique; they include the use of confocal laser scanning microscopy, digital image processing techniques (which hold promise for quantification of non-isotopic hybridization signal) and automated systems for non-isotopic *in situ* hybridization which may facilitate "routine" laboratory diagnosis of some viral infections

(Unger and Brigati, 1989). The use of the polymerase chain reaction "*in situ*" (Haase *et al.*, 1990) is an especially exciting development, which could be of considerable benefit in studies of latent viral infections.

The broad applicability and relative simplicity of the *in situ* hybridization technique have resulted in its adoption by an increasing number of researchers. A word of caution should be expressed however. The detection of viral sequences in a specimen does not necessarily indicate that the virus is responsible for the disease process evident. As methods of viral detection improve it is becoming increasingly apparent that we are all host to a large population of indigenous viral flora. The challenge lies in determining which of these viral genomes have pathological significance, and how the individual's response to the virus determines the eventual outcome of infection. It is in answering these questions (rather than merely detecting the presence of viral nucleic acid) that *in situ* hybridization has a unique advantage, allowing as it does the correlation of viral presence with morphological and functional changes in the infected cells.

ACKNOWLEDGEMENTS

The major contributions of Drs Helen Porter and Adam Padel to several aspects of the work discussed in this chapter are acknowledged, as is the excellent technical assistance provided by Clare Riley, Mark Evans, Angela Quantrill and Andrew Heyret. Probes were generously supplied by Prof H. zur Hausen, Heidelberg (HPV6), Dr P. Tattersall, Yale (pTY104), Dr H. Cooke, Edinburgh (pHY2.1) and Prof P. Griffiths, London (CMV HindIII fragment). Monoclonal antibodies were supplied by Dr D. Mason's laboratory, Oxford, and gold conjugates by Biocell. This work was supported by grants from the Wellcome Trust, the Oxford Regional Health Authority and the British Heart Foundation. ALM is a Nuffield Medical Research Fellow and Junior Research Fellow of New College, Oxford.

REFERENCES

Aksamit AJ, Mourrain P, Sever JL and Major EO (1985) Progressive multifocal leucoencephalopathy: investigation of three cases using *in situ* hybridization with JC virus biotinylated probe. Ann. Neurol. **18**: 490–496.

Alexandersen S, Bloom ME, Wolfinbarger J and Race RE (1987) *In situ* molecular hybridization for detection of Aleutian mink disease parvovirus DNA by using strand-specific probes: identification of target cells for viral replication in cell cultures and in mink kits with virus-induced interstitial pneumonia. J Virol. **61**: 2407–2419.

Ballinger ME, Rice CM and Miller BR (1988) Detection of yellow fever virus nucleic acid in infected mosquitoes by RNA:RNA *in situ* hybridization. Mol. Cell. Probes **2**: 331–338.

Bamborschke S, Porr A, Huber M and Heiss WD (1990) Demonstration of herpes simplex virus DNA in CSF cells by *in situ* hybridization for early diagnosis of herpes encephalitis. J. Neurol. **237**: 73–76.

Baskar JF, Stanat SC and Huang S (1986) Murine cytomegalovirus infection of mouse testes. J. Virol. **57**: 1149–1154.

Beckmann AM, Myerson D, Daling JR, Kiviat NB, Fenoglio CM and McDougall JK (1985) Detection and localization of human papillomavirus DNA in human genital condylomas by *in situ* hybridization with biotinylated probes. J. Med. Virol. **16**: 265–273.

Beckmann AM, Darling JR, Sherman KJ, Maden C, Miller BA, Coates RJ, Kiviat NB, Myerson D, Weiss NS and Hislop TG (1989) Human papilloma virus infection and anal cancer. Int. J. Cancer **43**: 1042–1049.

Benditt EP, Barrett T and McDougall JK (1983) Viruses in the etiology of atherosclerosis. Proc. Natl. Acad. Sci. USA **80**: 6386–6389.

Blum HE, Haase AT and Vyas GN (1984) Molecular pathogenesis of hepatitis B infection: simultaneous detection of viral DNA and antigens in paraffin-embedded liver sections. Lancet **2**: 771–775.

Bornkamm GW, Desgranges C and Gissman L (1983) Nucleic acid hybidization for the detection of viral genomes. Curr. Top. Microbiol. Immunol. **104**: 287–298.

Boyle AL, Ballard SG and Ward DC (1990) Differential distribution of long and short interspersed element sequences in the mouse genome: Chromosome karyotyping by fluorescence *in situ* hybridization. Proc. Natl. Acad. Sci. USA **87**: 7757–7761.

Brahic M, Stowring L, Ventura P and Haase AT (1981) Gene expression in visna virus infection in sheep. Nature **292**: 240–242.

Brahic M, Haase AT and Cash E (1984) Simultaneous *in situ* detection of viral RNA and antigens. Proc. Natl. Acad. Sci. USA **81**: 5445–5448.

Brahic M and Haase AT (1989) Double-label techniques of *in situ* hybridization and immunocyto-chemistry. Curr. Top. Microbiol. Immunol. 143: 9–20.

Brambilla C, Tackney C, Hirschman SZ, Colombo M, Dioguardi ML, Donato MF and Paronetto F (1986) Varying nuclear staining intensity of hepatitis B virus DNA in human hepatocellular carcinoma. Lab. Invest. **55**: 475–481.

Brigati DJ, Myerson D, Leary JJ, Spalholz B, Travis SZ, Fong CKY, Hsiung GD and Ward DC (1983) Detection of viral genomes in cultured cells and paraffin-embedded tissue sections using biotin-labelled hybidization probes. Virology **126**: 32–50.

Burns J, Graham AK, Frank C, Fleming KA, Evans MF and McGee J (1987) Detection of low copy human papilloma virus DNA and mRNA in routine paraffin sections of cervix by non-isotopic *in situ* hybridisation. J. Clin. Pathol. **40**: 858–864.

Burns J, Graham AK and McGee J (1988) Non-isotopic detection of *in situ* nucleic acid in cervix: an updated protocol. J. Clin. Pathol. **41**: 897–899.

Coghlan JP, Aldred P, Haralambidis J, Niall HD, Penschow JD and Tregear GW (1985) Hybridization Histochemistry. Analyt. Biochem. **149**: 1–28.

Cooke HJ, Schmidtke T and Gosden JR (1982) Characterisation of a human Y chromosome repeated sequence and related sequences in higher primates. Chromosoma **87**: 491–502.

Cordell JL, Fallini B, Erber WN, Ghosh AN, Abdulaziz Z, MacDonald S, Pulford KAF, Stein H and Mason DY (1984) Immunoenzymatic labelling of monoclonal antibodies using immune complexes of alkaline phosphatase and monoclonal antialkaline phosphatase. J. Histochem. Cytochem. **32**: 219–229.

Croen KD, Ostrove JM, Dragovic LJ and Straus SE (1988) Patterns of gene expression and sites of latency in human nerve ganglia are different for varicella-zoster and herpes simplex viruses. Proc. Natl. Acad. Sci. USA 85: 9773–9777.

Crum CP, Nuovo G, Friedman D and Silverstein SJ (1988) Accumulation of RNA homologous to human papillomavirus type 16 open reading frames in genital precancers. J. Virol. **62**: 84–90.

Crum CP and Roche JK (1990) Molecular pathology of the lower female genital tract. The papillomavirus model. Am. J. Surg. Pathol. **14** (Suppl 1): 26–33.

Cubie HA and Norval M (1989) Detection of human papilloma viruses in paraffin wax sections with biotinylated synthetic oligonucleotide probes and immunogold staining. J. Clin. Pathol. **42**: 988–991.

Eilbott DJ, Peress N, Burger H, LaNeve D, Ornstein J, Gendelman HE, Seidman R and Weiser B (1989) Human immunodeficiency virus tupe 1 in spinal cords of acquired immunodeficiency patients with myelopathy: expression and replication in macrophages. Proc. Natl. Acad. Sci. USA **86**: 3337–3341.

Fleming KA (1987) In-situ hybridization – a role in clinical pathology (Editorial). J. Pathol. **153**: 201–202.

Fleming KA, Evans M, Riley C, Franklin D, Morey A and Lovell-Badge R (1991) High sensitivity of non-isotopic *in situ* hybridization using digoxigenin labelled probes and transgenic mice. J. Pathol. **163**: 154A.

Fox JD, Briggs M, Ward PA and Tedder RS (1990) Human herpesvirus 6 in salivary glands. Lancet **336**: 590–593.

Gendelman HE, Narayan O, Stoskopf-Kennedy S, Kennedy PG, Ghotbi Z, Clements JE, Stanley J and Pezeshkpour G (1986) Tropism of sheep lentiviruses for monocytes: susceptibility to infection and virus gene expression increase during maturation of monocytes to macrophages. J. Virol. **58**: 67–74.

Gendelmen HE, Moench TR, Narayan O, Griffin DE and Clements JE (1985) A double label technique for performing simultaneous immunocytochemistry and *in situ* hybridization in virus infected cell cultures and tissues. J. Virol. Meth. **11**: 93–103.

Goldfarb J (1989) Leads from the MMWR. Risks associated with parvovirus B19 infection. J. Am. Med. Assoc. **261**: 1406–8 1555, 1560, 1563.

Grody WW, Cheng LS and Lewin KJ (1987) *In situ* viral DNA hybridization in diagnostic surgical pathology. Hum. Pathol. **18**: 535–543.

Grody WW, Lewin KJ and Naeim F (1988) Detection of cytomegalovirus DNA in classic and epidemic Karposi's sarcoma by *in situ* hybridization. Hum. Pathol. **19**: 524–528.

Haase AT, Stowring L, Ventura P, Burks J, Ebers G, Tourtelotte W and Warren K (1984) Detection by hybridization of viral infection of the human central nervous system. Ann. NY Acad. Sci. **436**: 103–108.

Haase AT, Walker D, Stowring L, Ventura P, Geballe A, Blum H, Brahic M, Goldberg R and O'Brien K (1985a) Detection of two viral genomes in single cells by double-label hybridization *in situ* and color microradioautography. Science **227**: 189–191.

Haase A, Gantz D, Eble B, Walker D, Stowring L, Ventura P, Blum H, Wietgrefe S, Zupancic M, Tourtellotte W, Gibbs C J, Norrby E and Rosenblatt S (1985b) Natural history of restricted synthesis and expression of measles virus genes in subacute sclerosing panencephalitis. Proc. Natl. Acad. Sci. USA **82**: 3020–3024.

Haase AT, Gantz D, Blum H, Stowring L, Ventura P, Geballe A, Moyer B and Brahic M (1985c) Combined macroscopic and microscopic detection of viral genes in tissues. Virology **140**: 201–206.

Haase A (1986) Analysis of viral infections by *in situ* hybridization. J. Histochem. Cytochem. **34**: 27–32.
Haase AT, Retzel EF and Staskus KA (1990) Amplification and detection of lentiviral DNA inside cells. Proc. Natl. Acad. Sci. USA **87**: 4971–4975.
Hadchouel M, Pasquinelli C, Fournier JG, Hugon RN, Scotto J, Bernard O and Brechot C (1988) Detection of mononuclear cells expressing hepatitis B virus in peripheral blood from HBsAg positive and negative patients by *in situ* hybridisation. J. Med. Virol. 24: 27–32.
Hamilton-Dutoit S, Pallesen G, Karkov J, Skinhoj P, Franzmann MB and Pedersen C (1989) Identification of EBV-DNA in tumour cells of AIDS-related lymphomas by in-situ hybridisation [letter]. Lancet **1**: 554–555.
Harders J, Lukacs N, Robert-Nicoud M, Jovin TM and Riesner D (1989) Imaging of viroids in nuclei from tomato leaf tissue by *in situ* hybridization and confocal laser scanning microscopy. EMBO J. **8**: 3941–3949.
Hawkins EP, Krischer JP, Smith BE, Hawkins HK and Finegold MJ (1990) Nasopharyngeal carcinoma in children – a retrospective review and demonstration of Epstein-Barr viral genomes in tumor cell cytoplasm: a report of the Pediatric Oncology Group. Hum. Pathol. **21**: 805–810.
Henderson A, Ripley S, Heller M and Kieff E (1983) Chromosome site for Epstein-Barr virus DNA in a Burkitt tumor cell line and in lymphocytes growth transformed *in vitro*. Proc. Natl. Acad. Sci. USA **80**: 1987–1991.
Herrington CS, Burns J, Graham AK, Bhatt B and McGee JO'D (1989a) Interphase cytogenetics using biotin and digoxigenin labelled probes II: simultaneous differential detection of two nucleic acid species in individual nuclei. J. Clin. Pathol. **42**: 601–606.
Herrington CS, Graham AK, Flannery DMJ, Burns J and McGee JO'D (1990) Discrimination of closely homologous HPV types by nonisotopic *in situ* hybridization: definition and derivation of tissue melting temperatures. Histochem. J. **22**: 545–554.
Hofler H (1987) What's new in "*in situ* hybridization." Pathol. Res. Pract. **182**: 421–430.
Hogg JC, Irving WL, Porter H, Evans M, Dunnill MS and Fleming K (1989) *In situ* hybridization studies of adenoviral infections of the lung and their relationship to follicular bronchiectasis. Am. Rev. Respir. Dis. **139**: 1531–1535.
Jilbert AR, Freiman JS, Gowans EJ, Holmes M, Cossart YE and Burrell C (1987) Duck hepatitis B virus DNA in liver, spleen, and pancreas: analysis by *in situ* and Southern blot hybridization. Virology **158**: 330–338.
Jones JF, Shurin S, Abramowsky C, Tubbs RR, Sciotto CG, Wahl R, Sands J, Gottman D, Katz BZ and Sklar J (1988) T-cell lymphomas containing Epstein-Barr viral DNA in patients with chronic Epstein-Barr virus infections. N. Engl. J. Med. **318**: 733–741.
Kennedy PG, Adams JH, Graham DI and Clements G (1988) A clinico-pathological study of herpes simplex encephalitis. Neuropathol. Appl. Neurobiol. **14**: 395–415.
Koenig S, Gendelman HE, Orenstein JM, Canto MCD, Pezeshkpour GH, Yungbluth M, Janotta F, Aksamit A, Martin MA and Fauci A (1986) Detection of AIDS virus in macrophages in brain tissue from AIDS patients with encephalopathy. Science **233**: 1089–1093.
Lawrence JB, Villnave CA and Singer RH (1988) Sensitive, high-resolution chromatin and chromosome mapping *in situ*: presence and orientation of two closely integrated copies of EBV in a lymphoma line. Cell **52**: 51–61.
Lawrence JB, Singer RH and Marselle LM (1989) Highly localized tracks of specific

transcripts within interphase nuclei visualized by *in situ* hybridization. Cell **57**: 493–502.

Lawrence JB, Marselle LM, Byron KS, Johnson CV, Sullivan JL and Singer RH (1990) Subcellular localization of low-abundance human immunodeficiency virus nucleic acid sequences visualized by fluorescence *in situ* hybridization. Proc. Natl. Acad. Sci. USA **87**: 5420–5424.

Lewandowski G, Delgado G, Holloway RW, Farrell M, Jenson AB and Lancaster WD (1990) The use of *in situ* hybridization to show human papillomavirus deoxyribonucleic acid in metastatic cancer cells within lymph nodes. Am. J. Obstet. Gynecol. **163**: 1333–1337.

Lewis SH, Reynolds-Kohler C, Fox HE and Nelson JA (1990) HIV-1 in trophoblastic and villous Hofbauer cells, and haematological precursors in eight-week fetuses. Lancet **335**: 565–568.

Lipkin WI, Villarreal LP and Oldstone MBA (1989) Whole animal section *in situ* hybridization and protein blotting: new tools in molecular analysis of animal models for human disease. Curr. Top. Microbiol. Immunol. **143**: 33–54.

Lo Y-MD, Mehal WZ and Fleming KA (1988) Rapid production of vector-free biotinylated probes using the polymerase chain reaction. Nuc. Acids Res. **16**: 8719.

Lohr JM and Oldstone MBA (1990) Detection of cytomegalovirus nucleic acid sequences in pancreas in type 2 diabetes. Lancet **336**: 644–648.

Maitland NJ, Kinross JH, Busuttil A, Ludgate SM, Smart GE and Jones KW (1981) The detection of DNA tumor virus-specific RNA sequences in abnormal human cervical biopsies by *in situ* hybridization. J. Gen. Virol. **55**: 123–137.

Maitland NJ, Cox MF, Lynas C, Prime S, Crane I and Scully C (1987) Nucleic acid probes in the study of latent viral disease. J. Oral. Pathol. **16**: 199–211.

Masih AS, Linder JL, Shaw BWJr, Wood RP, Donovan JP, White R and Markin RS (1988) Rapid identification of cytomegalovirus in liver allograft biopsies by *in situ* hybridization. Am. J. Surg. Pathol. **12**: 362–367.

McDougall JK, Dunn AR and Jones KW (1972) *In situ* hybridization of adenovirus RNA and DNA. Nature **236**: 346.

McDougall JK, Fenoglio CM and Galloway DA (1980) Cervical carcinoma: detection of herpes simplex virus RNA in cells undergoing neoplastic change. Int. J. Cancer **25**: 1–8.

McDougall JK, Myerson D and Beckmann AM (1986) Detection of viral DNA and RNA by *in situ* hybridization. J. Histochem. Cytochem. **34**: 33–38.

Milde K and Loning T (1986) Detection of papillomavirus DNA in oral papillomas and carcinomas: application of *in situ* hybridization with biotinylated HPV 16 probes. J. Oral. Pathol. **15**: 292–296.

Moen PTJr, Fox E and Bodnar JW (1990) Adenovirus and minute virus of mice DNAs are localized at the nuclear periphery. Nuc. Acids Res. **18**: 513–519.

Moench TR (1987) *In situ* hybridization (review). Mol. Cell. Probes **1**: 195–205.

Morey AL, Fleming KA, del-Buono R and Chandler JA (1991a) A flexible method for non-isotopic *in situ* labelling of multiple nucleic acid and antigenic targets in individual cells or sections. J. Pathol. **163**: 159A.

Morey L, Fleming KA, Keeling JW and Porter HJ (1991b) Diagnosis and investigation of human fetal parvovirus infection by *in situ* hybridization combined with immunophenotyping of infected cells (submitted for publication)..

Morey AL, Fleming KA, Ferguson D and Sutton L (1991c) Cellular features of parvovirus infection *in vitro*. J. Pathol. **163**: 168A.

Mullink H, Walboomers JM , Raap AK and Meyer CJ (1989a) Two colour DNA *in situ* hybridization for the detection of two viral genomes using non-radioactive probes. Histochemistry **91**: 195–198.

Mullink H, Walboomers JMM, Tadema TM, Jansen D and Meijer CJLM (1989b) Combined immuno- and non-radioactive hybridocytochemistry on cells and tissue sections: influence of fixation, enzyme pretreatment, and choice of chromogen on detection of antigen and DNA sequences. J. Histochem. Cytochem. 37: 603–609.

Myerson D, Hackman RC, Nelson JA, Ward DC and McDougall JK (1984) Widespread presence of histologically occult cytomegalovirus. Hum. Pathol. **15**: 430–439.

Myerson D (1988) *In Situ* Hybridizaton. In: Diagnostic Immunopathology. (R. B. Colvin, A. K. Bhan and R. T. McCluskey eds.),pp 475–498 New York: Raven Press.

Naoumov NV, Alexander GJ, Eddleston AL and Williams R (1988a) *In situ* hybridisation in formalin fixed, paraffin wax embedded liver specimens: method for detecting human and viral DNA using biotinylated probes. J. Clin. Pathol. **41**: 793–798.

Naoumov NV, Alexander GJM, O'Grady JG, Aldis P, Portman BC and Williams R (1988b) Rapid diagnosis of cytomegalovirus infection by in-situ hybridisation in liver grafts. Lancet **1**: 1361–1364.

Negro F, Bonino F, di Bisceglie A, Hoofnagle JH and Gerin JL (1989) Intrahepatic markers of hepatitis delta virus infection: a study by *in situ* hybridization. Hepatology **10**: 916–920.

Norval M and Bingham RW (1987) Advances in the use of nucleic acid probes in diagnosis of viral diseases of man. Arch. Virol. **97**: 151–165.

Orth G, Jeanteur P and Croissant O (1970) Evidence for and localization of vegetative viral DNA replication by autoradiographic detection of RNA-DNA hybrids in sections of tumors induced by Shope papilloma virus. Proc. Natl. Acad. Sci. USA **68**: 1876–1880.

Ozden S, Aubert C, Gonzalez-Dunia D and Brahic M (1990) Simultaneous *in situ* detection of two mRNAs in the same cell using riboprobes labeled with biotin and ^{35}S. J. Histochem. Cytochem. **38**: 917–922.

Padel AF, Venning VA, Evans MF, Quantrill AM and Fleming KA (1990) Human papillomaviruses in anogenital warts in children: typing by *in situ* hybridisation. Br. Med. J. **300**: 1491–1494.

Patel D, Shepherd PS, Naylor JA and McCance DJ (1989) Reactivities of polyclonal and monclonal antibodies raised to the major capsid proteins of human papillomavirus type 16. J. Gen. Virol. **70**: 69–77.

Peluso R, Haase AT, Stowring L, Edwards M and Ventura P (1985) A Trojan Horse mechanism for the spread of visna virus in monocytes. Virology **147**: 231–236.

Perlman S, Jacobsen G and Afifi A (1989) Spread of a neurotropic murine coronavirus into the CNS via the trigeminal and olfactory nerves. Virology **170**: 556–560.

Pezzella M, Rossi P, Lombardi V, Gemelli V, Mariani-Costantini R, Mirolo M, Funaro C, Moschese V and Wigzell H (1989) HIV viral sequences in seronegative people at risk detected by *in situ* hybridization and polymerase chain reaction. Br. Med. J. **298**: 713–716.

Popescu NC and DiPaolo JA (1990) Integration of human papillomavirus 16 DNA and genomic rearrangements in immortalized human keratinocyte lines. Cancer Res. **50**: 1316–1323.

Porter HJ, Khong TY, Evans MF, Chan VT and Fleming KA (1988) Parvovirus as a cause of hydrops fetalis: detection by *in situ* DNA hybridisation. J. Clin. Pathol. **41**: 381–383.

Porter HJ, Heryet A, Quantrill AM and Fleming KA (1990) Combined non-isotopic in-situ hybridization and immunohistochemistry on routine paraffin wax embedded tissue: identification of cell type infected by human parvovirus and demonstration of cytomegalovirus DNA and antigen in renal infection. J. Clin. Pathol. **43**: 129–132.

Puvion-Dutilleul F and Puvion E (1989) Ultrastructural localization of viral DNA in thin sections of herpes simplex virus type 1 infected cells by *in situ* hybridization. Eur. J. Cell. Biol. **49**: 99–109.

Puvion-Dutilleul F, Pichard E, Laithier M and Puvion E (1989) Cytochemical study of the localization and organization of parental herpes simplex virus type I DNA during initial infection of the cell. Eur. J. Cell. Biol. **50**: 187–200.

Salimans MM, van de Rijke FM, Raap AK and van Elsacker-Niele AM (1989) Detection of parvovirus B19 DNA in fetal tissues by *in situ* hybridization and polymerase chain rection. J. Clin. Pathol. **42**: 525–530.

Stevens JG (1989) Herpes simplex virus latency analyzed by *in situ* hybridization. Curr. Top. Microbiol. Immunol. **143**: 1–8.

Stowring L, Haase AT, Petursson G, Georgsson G, Palsson P, Lutley R, Roos R and Szuchet S (1985) Detection of visna virus antigens and RNA in glial cells in foci of demyelination. Virology **141**: 311–318.

Syrjänen KJ (1987) Biology of human papillomavirus (HPV) infections and their role in squamous cell carcinogenesis. Med. Biol. **65**: 21–39.

Syrjänen S, Cintorino M, Armellini D, delVecchio MT, Leoncini P, Bugnoli M, Pallini V, Silvestri S, Tosi P, Mantyjarvi R, *et al.* (1988) Expression of cytokeratin polypeptides in human papillomavirus (HPV) lesions of the uterine cervix: 1. Relationship to grade of CIN and HPV type. Int. J. Gynecol. Pathol. **7**: 23–38.

Teo CG and Griffin B (1987) Epstein-Barr virus genomes in lymphoid cells: activation in mitosis and chromosomal location. Proc. Natl. Acad. Sci. USA **84**: 8473–8477.

Terry RM, Lewis FA, Robertson S, Blythe D and Wells M (1989) Juvenile and adult laryngeal papillomata: classification by in-situ hybridization for human papillomavirus. Clin. Otolaryngol. **14**: 135–139.

Turtinen LW, Saltzman R, Jordan MC and Haase A (1987) Interactions of human cytomegalovirus with leukocytes *in vivo*: analysis by *in situ* hybridization. Microb. Pathog. **3**: 287–297.

Unger ER and Brigati DJ (1989) Colorimetric in-situ hybridization in clinical virology: development of automated technology. Curr. Top. Microbiol. Immunol. 143: 21–31.

Vallejos H, del Mistro A, Kleinhaus S, Braunstein JD, Halwer M and Koss LG (1987) Characterization of human papilloma virus types in condylomata acuminata in children by *in situ* hybridization. Lab. Invest. **56**: 611–615.

van der Loos CM, Volkers HH, Rook R, van den Berg FM and Houthoff H-J (1989) Simultaneous application of *in situ* DNA hybridization and immunohistochemistry on one tissue section. Histochem. J. **21**: 279–284.

Vazeux R, Cumont M, Girard PM, Nassif X, Trotot P, Marche C, Matthiessen L, Vedrenne C, Mikol J, Henin D, *et al.* (1990) Severe encephalitis resulting from coinfections with HIV and JC virus. Neurology **40**: 944–948.

Venables PJ, Teo CG, Baboonian C, Griffin BE and Hughes R (1989) Persistence of Epstein-Barr virus in salivary gland biopsies from healthy individuals and patients with Sjogren's syndrome. Clin. Exp. Immunol. 75: 359–364.

Weiss LM, Movahed LA, Warnke RA and Sklar J (1989) Detection of Epstein-Barr viral genomes in Reed-Sternberg cells of Hodgkin's disease. N. Engl. J. Med. **320**: 502–506.
Wolber RA and Lloyd RV (1988) Cytomegalovirus detection by nonisotopic *in situ* DNA hybridization and viral antigen immunostaining using a two-color technique. Hum. Pathol. **19**: 736–741.
Wolber RA, Beals TF, Lloyd RV and Maassab H (1988) Ultrastructural localization of viral nucleic acid by *in situ* hybridization. Lab. Invest. **59**: 144–151.
Wolber RA, Beals TF and Maassab H (1989) Ultrastructural localization of herpes simplex virus RNA by *in situ* hybridization. J. Histochem. Cytochem. **37**: 97–104.
Wolf BC, Martin AW, Neiman RS, Janckila AJ, Yam LT, Caracansi A, Leav BA, Winpenny R, Schultz DS and Wolfe HJ (1990) The detection of Epstein-Barr virus in hairy cell leukemia cells by *in situ* hybridization. Am. J. Pathol. **136**: 717–723.
Yamashiroya HM, Ghosh L, Yang R and Robertson ALJ (1988) Herpesviridae in the coronary arteries and aorta of young trauma victims. Am. J. Pathol. **130**: 71–79.

5
In situ hybridization for molecular cytogenetics

A.K. Raap[1], C.J. Cornelisse[2]
Department of Cytochemistry and Cytometry[1], Medical Faculty, Leiden University, Wassenaarseweg 72, 2333 AL Leiden, The Netherlands. Department of Pathology[2], Medical Faculty, Leiden University, Wassenaarseweg 62, 2333 AL Leiden, The Netherlands

1. HISTORICAL PERSPECTIVE

After the biochemical description of the principles of nucleic acid hybridization in the sixties, three groups reported independently in 1969–1970 its application for the microscopic detection of specific nucleic acid sequences (Pardue and Gall, 1969; John *et al.*, 1969; Buongiorno-Nordelli and Amaldi, 1970). In contrast to the first microscopic work done with fluorescent antibody probes, the labels were radioisotopes, the detection of which was accomplished by micro-autoradiography. The disadvantages inherent to the use of radioisotopes (poor topological resolution, environmental and health hazards, complex multiple sequence detection) prompted several groups to develop non-isotopic nucleic acid detection techniques.

Modifications of nucleic acid probes for this purpose should not lead to loss of hybridization properties. Rudkin and Stollar (1977) bypassed this potential problem by employing anti-DNA:RNA antibodies to detect immunocytochemically specific RNA:DNA hybrids. They were the first to appreciate the sensitivity and high resolution of fluorescence *in situ* hybridization. Their method is, however, strongly dependent upon availability of high affinity and monospecific anti-DNA:RNA antisera, the production of which appears to be cumbersome (Van Prooijen-Knegt *et al.*, 1982; Kitigawa and Stollar, 1984; Raap *et al.*, 1984).

To disturb hybridization properties minimally, Bauman *et al.* (1980) developed a 3′-terminal fluorescent RNA labelling technique, based on periodate oxidation of RNA probes to create at the 3′-end aldehyde functions and aldehyde-reactive derivatives of the commonly used fluorochromes fluorescein and rhodamine. This direct method was limited in its sensitivity and, therefore, immunocytochemical amplification of signal was performed using anti-fluorescein antibodies (Schmitz and Kampa, 1979; Bauman *et al.*, 1981).

In the 1980s the principle of immunocytochemical detection of "hapten-modified" DNA probes has been explored much further resulting in a large number of hapten labelling procedures (for review of methods developed until 1988, see Raap *et al.*,1989: Raap *et al.*, 1990; Coulton, 1990). Of special interest in this respect is the reactivity of the C5 position of (deoxy)uridine for metal ions such as Hg^{2+} (Dale *et al.*, 1973). (d)UTP mercurated at the C5 position forms an important intermediate in the synthesis of allylamine-dUTP, which is the key compound for covalent attachment of biotin or digoxigenin, via their *N*-hydroxysuccinimide esters, to (d)UTP with proper spacing (Langer *et al.*,1981; Brigati *et al.*, 1982: see Chapter 1, Figure 4). DNA and RNA polymerases accept such modified DNA/RNA precursors. Therefore, this labelling format fits those of well established enzymatic methods for radioisotopic labelling of nucleic acids (i.e. nick translation, primer extension, terminal transferase reactions) and thus provides the tool for wide application of non-isotopic nucleic acid labelling techniques. In fact any hapten or fluorochrome that is reactive with primary amines can be used in this approach as long as the polymerases will accept the modified substrate with reasonable efficiency.

Next to these enzymic approaches of nucleic acid hapten modification, several chemical routes have been taken to accomplish the same. Relevant examples are the acetylaminofluorene (Landegent *et al.*, 1984; Tchen *et al.*, 1984), the mercury/sulphydryl-hapten (Hopman *et al.*, 1986), the transamination-haptenization (Sverdlov *et al.*, 1974; Viscidi *et al.*, 1986), and the photoactivated hapten (Forster *et al.*, 1985; Keller *et al.*, 1989) procedures. General advantages of these chemical labelling procedures are their ease of performance and the fact that relatively large amounts of probe can be produced in one reaction.

In the ten years or so of their existence, the non-isotopic *in situ* hybridization techniques have fully profited from half a century of continuous improvements of immunocytochemistry and microscopy. Major achievements in immunocytochemistry in this respect are the monoclonal antibody technology, the chemically well defined conjugation of reporter molecules like fluorochromes, enzymes and colloidal gold, sharply localizing enzyme cytochemical methods, the discovery of the biotin-avidin system, immunological amplification techniques as well as silver amplification techniques for gold and enzyme reporters. In the field of microscopy, the development of epifluorescence, reflection contrast and laser scanning (confocal) microscopy has made significant contributions. Clearly, also the major developments in recombinant DNA technology and, more recently, *in vitro* chemical as well as enzymatic DNA synthesis played an important role.

Through the combined efforts of several research groups *in situ* hybridization techniques have come to the stage that multiple single copy DNA sequences as well as (pre)mRNAs can be detected in multiple colours in individual cells (Landegent *et al.*, 1985; Hopman *et al.*, 1986; Pinkel *et al.*, 1986; Lawrence *et al.*, 1988; Lawrence *et al.*, 1989; Lawrence *et al.*, 1990; Bhatt *et al.*, 1988; Cherif *et al.*, 1989; Nederlof *et al.*, 1989a; Nederlof *et al.*, 1990; Lichter *et al.*, 1990; Dirks *et al.*, 1989; Dirks *et al.*, 1990; Wiegant *et al.*, in press).

Table 1 Current status of non-isotopic *in situ* hybridization

sensitivity:	unique DNA sequences of 1–5 kb
resolution:	metaphase ≥ 1000 kbp
	interphase ≥ 100 kbp
multiplicity:	≥ 4
detection of (pre) mRNA	
combination with immunocytochemistry	

Also the combination of *in situ* hybridization and immunocytochemistry is feasible. In Table 1 we summarize the status of *in situ* hybridization as of October 1990.

2. PROSPECTS

It is clear from the above that with the current *in situ* hybridization technology the number of applications are already numerous. Nevertheless, the techniques have not been developed and investigated to their utmost.

First, a re-exploration of the direct nucleic acid labelling techniques is currently being made, because such techniques will contribute to simplification and speed of the methods. Preliminary results from *in situ* hybridization experiments with fluorochrome labelled DNA in our laboratory indicate that the sensitivity that can be reached by visual assessment of results is in the order of 50–100 kbp unique DNA. Although this sensitivity is a factor of about ten less than that attainable by indirect techniques, these results imply that, in the near future, molecular cytogenetics with probes spanning a few 100 kb of DNA will be possible with such directly fluorochromized probes . With the advent of yeast artificial chromosome (YAC) cloning and the ease of current cosmid/phage cloning techniques, the production of such contiguous DNA may be considered a routine task. These experiments with fluorochromized probes proved furthermore that the background signals frequently observed with indirect techniques are mainly due to the immunocytochemical detection steps, thereby partly explaining the relative good sensitivity of the direct approach. Secondly, the application of advanced opto-electronics for sensitive imaging, image processing and evaluating of *in situ* hybridization results is still in its infancy (Tanke, 1989). With implementation of microscopic imaging using new generation, sensitive and integrating solid state cameras free of image distortion, it may be expected that sensitivity of detection will increase even further. Because of the low background they provide, direct techniques are of particular interest in this respect.

Also for increasing multiplicity of *in situ* hybridization, sophisticated image cytometry of ISH results is a prerequisite. In this respect it is useful to recall that currently the multiplicity of *in situ* hybridization is limited by the number of immunofluorophores available and not the number of DNA-hapten modifications. In fact only red, green and blue fluorescent colours are in practice available, implying that the number of simultaneously detectable probes is only three (Nederlof *et al.,* 1989: Colour plate 6). However, by using double labelled probes the multiplicity can be increased (see Table 2).

For instance, in a quadruple *in situ* hybridization three of the four probes would be identified on basis of their unique colour and the fourth on the basis of its double fluorescence colour, provided the hybridization signals do not overlap.

However, visual examination becomes tedious and fluorescence ratio imaging with pseudo-colour display would be of help in such a situation. Recently, we have analysed such double labelled probes by digital fluorescence image cytometry (Nederlof *et al.*, unpublished results). The results suggest strongly,

Table 2 Principle of multiple fluorescence *in situ* hybridization

	fluo-green	*dig-red*	*bio-blue*
probe-1-fluo	+	–	–
probe-2-dig	–	+	–
probe-3-bio	–	–	+
probe-4-fluo-dig	+	+	–
probe-5-fluo-bio	+	–	+
probe-6-dig-bio	–	+	+
probe-7-fluo-dig-bio	+	+	+

Explanation: probes are single, double or triple labelled with digoxogenin (dig), biotin (bio) or fluorescein (fluo). After *in situ* hybridization the biotin moiety is detected in blue fluorescence (e.g. from amino methyl coumarin acetic acid, AMCA), the digoxigenin in red fluorescence (e.g. from Texas Red) and the fluorescein moiety is detected in green fluorescence either directly or indirectly. According to this principle, a maximum of seven different probes can be detected (see Nederlof *et al.*, 1990).

that the ratios of the fluorescence intensities derived from double haptenized probes are fairly constant (typical coefficients of variation are 10–15%), implying that by using different ratios of DNA labelling (direct or indirect) and digital fluorescence image cytometry, the multiplicity in principle can be increased even further. Speed of analysis of *in situ* hybridization is another important issue where image cytometry can have major contributions. For instance, in "interphase cytogenetics" (see below) often many nuclei have to be evaluated for single or multiple *in situ* hybridization signals. For humans such a task is boring and development of automated microscopic instrumentation and image processing is warranted for this purpose.

In conclusion, *in situ* hybridization has already reached a high level of sophistication, but has not yet reached its methodological endpoint.

3. APPLICATION TO GENE MAPPING

In gene mapping and genetic linkage studies it is frequently of importance to rapidly map newly cloned DNA sequences on chromosomes. Because it is the most direct technique and as it has the required sensitivity *in situ* hybridization is exquisitely suited for this purpose (Landegent *et al.*, 1985, 1987). Some recent examples of application of ISH in DNA mapping can be found in Dauwerse *et al.* (1989); Kievits *et al.* (1990); Lichter *et al.* (1990) and Wiegant *et al.* (in press).

For a correct cytogenetic assignment, the counterstain should have sufficient banding information to allow accurate classification of the chromosome(s) bearing the *in situ* hybridization signal. In our experience a DAPI/chromomycine combination suffices mostly. Other examples of a procedure combining *in situ* hybridization and chromosome banding has recently been presented by Cherif *et al.* (1990) and Fan *et al.* (1990). Lichter *et al.* (1990) uses a digital imaging technique and fractional length measurements for mapping purposes. To take away any ambiguity in chromosomal assignment double fluorescence *in situ* hybridization with the specific probe and chromosome DNA libraries (chromosome painting) is of particular importance as recently shown by Wiegant *et al.* (1992, in press).

Of particular interest in gene mapping is the resolution of the *in situ* hybridization technique in terms of physical DNA distance. Elegant work in this respect has been done by Trask *et al.* (1989) and Lawrence *et al.*, (1990) who demonstrated that in interphase nuclei the lower limit of resolution of uni-colour ISH can be less than 100 kb (i.e. a distance which can be spanned by 2–3 cosmids). Because of the compaction, in metaphase the resolution is in the order of several Mb. The multiple ISH techniques are of particular interest for rapid gene mapping purposes, because they will contribute to increasing the speed with which newly generated DNA clones can be mapped.

4. APPLICATIONS IN TUMOUR CYTOGENETICS

It is well established that chromosomal aberrations play a significant role in tumourigenesis. For instance, cytogenetic analysis of leukaemias and lymphomas has revealed many tumour-specific chromosomal abnormalities. Although a growing number of consistent cytogenetic abnormalities in both malignant and benign solid tumours are now being revealed (see Heim and Mitelman, 1987) it has not been possible to cytogenetically analyse solid tumours on the scale that is possible for haematological malignancies. This is the consequence of a number of technical difficulties (for review see Teyssier, 1989):

(i) the mitotic index of most solid tumours.

(ii) in most cases a more or less extensive culturing is necessary.

(iii) such culturing may lead to selective overgrowth by subpopulations of cells, among which might be non-neoplastic cells and also cell culture itself may induce changes in karyotype like e.g. tetraploidization.

(iv) metaphase spreads from solid tumours are often of inferior quality and yield fuzzy chromosomes with poor banding patterns, making correct classification extremely difficult.

(v) solid tumours often show a complex karyotype with numerous abnormalities, which probably reflects the karyotypic evolution, but obscures primary events.

In situ hybridization with a variety of DNA probes can be of great help in improving the classification of metaphase chromosomes from solid tumour cells. A recent study by Smit *et al.* (1990) may illustrate this point. Using a combined Giemsa banding–*in situ* hybridization approach, they showed for an ovarian cancer cell line that the cytogenetically assigned Chromosome 8p– actually contained the pericentromeric region of Chromosome 6. A similar observation was made for the cytogenetically assigned Chromosome 10p+. Furthermore, the techniques allowed a better identification of one of the marker chromosomes, which showed hybridization signals with probes for the pericentromeric regions of Chromosome 1 and Chromosome 10. It is within reach to analyse complex translocations by using DNA probe (cocktail)s specific for parts of or entire chromosomes and multiple fluorescence *in situ* hybridization (Nederlof *et al.*, 1989a; 1990). It is of significant importance to remark that it is also possible to analyse the hybridization signals in interphase nuclei, with or without intermittent culturing. This approach, for which Cremer *et al.,* (1986) proposed the term "interphase cytogenetics" takes away some, but not all of the technical difficulties mentioned above. A great advantage of interphase cytogenetics is that it does allow a very rapid enumeration of *in situ* hybridization signals per cell and, therefore, in principle overcomes the problem of representation of cell subpopulations (Colour plate 7). For instance, the counting of hybridization signals of 200–400 nuclei takes only 10–20 minutes. However, extrapolation of the data in terms of numerical chromosome aberrations is not legitimate, because in fact only a very small portion of the genome is being analysed.

Several studies have been undertaken to validate the interphase cytogenetic approach. These show both its major advantage (rapid analysis of many nuclei) and disadvantage (only limited information is obtained in one hybridization, which cannot *a priori* be interpreted cytogenetically). For instance, in a blind study Nederlof *et al.* (1989b), performed *in situ* hybridization to interphase nuclei of several tumours with a #1satellite DNA probe and found good correlation with the karyotypic data. However, only in retrospect a translocation between Chromosome 1 and Chromosome 7 could be identified in interphase. Though multi-colour simultaneous hybridization can be used to solve this problem partially, it is clear that knowledge of the karyotypes is desirable in interphase cytogenetics. Therefore, interphase cytogenetics performs best in those situations where prototypic abnormalities are known. This is nicely illustrated by recent work of Arnoldus *et al.* (1990a), who describe the detection of the well known t(9,22) translocation (Philadelphia chromosome) in interphase nuclei of patients with chronic myeloid leukaemia. The translocation brings sequences from the abl oncogene on Chromosome 9 in close apposition to bcr gene sequences on Chromosome 22. The detection principle is based on the fact that abl and bcr sequences, when hybridized *in situ* in different fluorescent colours, will appear in the microscope as two colocalizing spots, while the abl and bcr signals on the normal Chromosome 9 and Chromosome 22 homologues, respectively, appear as separate spots. In normal cells two times two separate spots are seen.

Thus, to monitor patients for residual cancer cells of which one or more cytogenetic hallmarks are known, interphase cytogenetics is exquisitely suited. As shown for the Philadelphia chromosome the aberrations detectable in interphase by *in situ* hybridization, include structural ones such as translocations. The better statistics obtainable by interphase cytogenetics may help to validate the significance of, for example, trisomies found in only a few metaphases of patients as recently shown by Kibbelaar *et al.* (1990) for trisomy 8 in patients with myelodysplastic syndrome. As already stated, much more well established knowledge exists of chromosome aberration in haematological malignancies than in solid tumours. Therefore, interphase cytogenetics will probably be implemented rapidly for diagnosis and monitoring of leukaemias and lymphomas.

For solid tumours the prognostic value of (flow) cytometric nuclear DNA content measurements has been widely studied (Cornelisse and Tanke, 1990). In contrast to cytogenetics using banding, flow cytometry can be used routinely on a large scale and yields data representative of the entire cell population. Although DNA cytometry is seemingly accurate (2–4% coefficients of variations are routinely obtainable with clinical samples), it provides no information about the involvement of specific chromosomes. Also, its resolution is insufficient to detect in heterogeneous clinical samples DNA content differences of less than 4% i.e. the loss or gain of one average-sized chromosome (Vindelov *et al.*, 1983; Raap *et al.*,1987). This illustrates that measurement of total nuclear DNA content by flow cytometry, although providing global tumour-ploidy information, is inappropriate for the study of chromosomal aberrations.

In view of this, and the enormous logistic and technical difficulties encountered in conventional cytogenetic studies of solid tumours (for review, see Teyssier, 1989) several groups have started to develop and apply interphase cytogenetics in pathology in spite of back-up from accurate karyotypes and lack of prototypic abnormalities. A recent study by Hopman *et al.* (1989), may illustrate that a much more refined analysis of individual genomes is possible with *in situ* hybridization than with DNA flow cytometry. For instance, in a series of bladder tumours with a DNA index of 1 (implying a normal DNA content), they found that 25% of the tumours had numerical aberrations. Similar results have been reported for breast cancer by Devilee *et al.* (1988). It should be emphasized again, however, that due to the often complex karyotypes shown by solid tumours, extrapolation of interphase cytogenetic data towards the karyotype level is not without pitfalls. For example, due to complex structural rearrangements, the number of signals obtained for a single chromosome-specific probe may not reflect the true copy number of the intact cognate chromosome.

Recently, we have shown, that technical developments now allow for the preparation of suspensions of solid tumour nuclei that are amenable to double fluorescence *in situ* hybridization at the single copy gene level (Arnoldus *et al.*, in press). Furthermore, more knowledge of the molecular genetic and chromosomal basis of the various solid tumours will emerge.

Thus it seems reasonable to assume that interphase cytogenetics of solid tumours will eventually be performed with probes for genes that play a pathophysiological role in the tumour origin and progression. In this way, we see a role for interphase cytogenetics that complements molecular genetic studies of genomic changes in tumours based on the analysis of restriction fragment length polymorphisms (RFLP). Interphase cytogenetics, by providing information on single gene loci in individual cells, could greatly contribute to our insight into genetic tumour cell heterogeneity. Such information is largely lost in DNA samples obtained by biochemical extraction of tumour tissue. Thus, for well-defined chromosomal regions, the combination of interphase cytogenetics and RFLP analysis may well turn out to be a more sensitive as well as more generally applicable approach to the cytogenetic study of solid tumours than conventional karyotype analysis. As a consequence of the current and future developments in interphase cytogenetics, new problems will arise. One example is reflected by the findings of Arnoldus *et al.* (1989, 1990b) that in normal cells somatic pairing of centromeres may occur, resulting in one hybridization spot, which might erroneously be interpreted as a monosomy. Another problem that is already evident now, relates to the fact that microscopic evaluation of *in situ* hybridization results of many nuclei is fatiguing. A remedy for this will have to be found in the development of automated microscopy systems capable of rapidly analysing large number of cells labelled with multiple colours .

5. CONCLUSIONS

In the relative short time of their existence, the non-isotopic *in situ* hybridization techniques have developed into widely applicable research tools, but they have not yet reached their methodological endpoint. For DNA mapping purposes they will be widely applied. In the field of cytogenetics they will:

(i) assist in more accurate metaphase chromosome classification.

(ii) provide powerful tools for diagnosis and monitoring tumours with known chromosomal changes.

(iii) contribute significantly to refinement of diagnosis and prognosis of tumours with unknown chromosome abnormalities.

REFERENCES

Arnoldus EPJ, Peters ACB, Bots GTAM, Raap AK and Van der Ploeg M (1989) Somatic pairing of chromosome 1 centromeres in interphase nuclei of human cerebellum. Hum. Genet. **83**: 231–234.

Arnoldus EPJ, Wiegant J, Noordermeer IA, Wessels JW, Beverstock GC, Grosveld GC, Van der Ploeg M and Raap AK (1990) Detection of the Philadelphia chromosome in interphase nuclei. Cytogenet. Cell Genet. **54**: (3–4): 108–111.

Arnoldus EPJ, Noordermeer IA, Peters ACB, Voormolen JHC, Bots GTAM, Raap AK and

Van der Ploeg M (1991) Interphase cytogenetics of brain tumors. Genes, Chromosomes and Cancers. In press.

Arnoldus EPJ, Noordermeer IA, Peters ACB, Raap AK and Van der Ploeg M (1990) Interphase cytogenetics reveals somatic pairingof chromosome 17 centromeres in normal human brain, but no trisomy 7 or sex-chromosome loss. Cytogenet. Cell Genet. In press .

Bauman JGJ, Wiegant J, Borst P and Van Duijn P (1980) A new method for fluorescence microscopical localization of specific DNA sequences by *in situ* hybridization of fluorochrome labeled RNA. Exp. Cell Res. **138**: 485–490.

Bauman JGJ, Wiegant J and Van Duijn P (1981) Cytochemical hybridization with fluorochrome labeled RNA III. Increased sensitivity by the use of anti-fluorescein antibodies. Histochemistry **73**: 181–193.

Bhatt B, Burns J, Flannery D and McGee JO'D (1988) Direct visualization of single copy genes on banded metaphase chromosomes by non-isotopic *in situ* hybridization. Nuc. Acids Res. **16**: 3951–3961.

Brigati DJ, Myerson D, Leary JJ, Spalholz B, Travis S, Fong CK, Hsiung GD and Ward DC (1982) Detection of viral genomes in cultured cells and paraffin embedded tissue sections using biotin labelled probes. Virology **126**: 32–50.

Buongiorno-Nordelli N and Amaldi F (1969) Autoradiographic detection of molecular hybrids between rRNA and DNA in tissue sections. Nature **2254**: 946–947.

Cherif D, Bernard O and Berger R (1989) Detection of single-copy genes by non-isotopic *in situ* hybridization on human chromosomes. Hum. Genet. **81**: 358–362.

Cherif D, Julier C, Delattre O, Derre J, Lathrop GM and Berger R (1990) Simultaneous localization of cosmids and chromosome R-banding by fluorescence microscopy: application to regional mapping of human chromosome 11. Proc. Natl. Acad. Sci. USA **87**: 6639–66438.

Cornelisse CJ and Tanke HJ (1990) Flow Cytometry. In: *Comprehensive Cytopathology*. (Bibbo M ed), Saunders, In Press.

Coulton G (1990) Non-radioisotopic labels for *in situ* hybridization histochemistry: a histochemists view. In: *In Situ Hybridization: Application to Developmental Biology and Medicine*. (Harris N and Wilkinson DG eds.), pp. 1–32, Cambridge University Press.

Cremer T, Landegent JE, Bruckner A, Scholl HP, Schardin M, Hager HD, Devilee P, Pearson P and Van der Ploeg M (1986) Detection of chromosome aberrations in the human interphase nucleus by visualization of specific target DNAs with radioactive and non-radioactive *in situ* hybridization techniques: diagnosis of trisomy 18 with probe L1.84. Hum. Genet. **74**: 346–352.

Dauwerse JG, Kievits T, Beverstock GC, Van der Keur D, Smit E, Wessels HW, Hagemeijer A, Pearson PL, Van Ommen GJB and Breuning MH (1990) Rapid detection of chromosome 16 inversion in acute nonlymphocytic leukemia, subtype M4: regional localization of the breakpoint in 16p. Cytogenet. Cell Genet. **53**: 126–12.

Devilee P, Thierry RF, Kolluri R, Hopman AHN, Willard HF, MHPearson PL and Cornelisse CJ (1988) Detection of chromosome aneuploidy in interphase nuclei from human primary breast tumours using chromosome specific repetitive DNA probes. Cancer Res. **48**: 5825–5830.

Dirks RW, Raap AK, Van Minnen J, Vreugdenhil E, Smit AB and Van der Ploeg M (1989) Detection of mRNA molecules coding for neuropeptide hormones of the pond snail Lymnaea stagnalis by radioactive and non-radioactive *in situ* hybridization: a model study for mRNA detection. J. Histochem. Cytochem. **37**: 7–14.

Dirks RW, Van Gijlswijk RPM, Tullis RH, Smit AB, Van Minnen J, Van der Ploeg M and Raap AK (1990) Simultaneous detection of different mRNA sequences coding for neuropeptide hormones by double *in situ* hybridization using FITC- and biotin-labeled oligonucleotides. J. Histochem. Cytochem. **38**: 467–473.

Dale RMK, Livingstone DC and Ward DC (1973) The synthesis and enzymatical polymerization of nucleotides containing mercury: potential tools for nucleic acid sequencing and structural analysis. Proc. Natl. Acad. Sci. USA **70**: 2238–2242.

Fan Y, Davis LM and Shows TB (1990) Mapping small DNA sequences by fluorescence *in situ* hybridization directly on banded metaphase chromosomes. Proc. Natl. Acad. Sci. USA **87**: 6223–6227.

Forster AC, McInnes JL, Skingle DC and Symons RH (1985) Non-radioactive hybridization probes prepared by the chemical labeling of DNA and RNA with a novel reagent: photobiotin. Nuc. Acids Res. **13**: 745–761.

Heim S, Mitelman F (1987) In: *Cancer Cytogenetics*, Alan R. Liss Inc, New York.

Hopman AHN, Wiegant J, Tesser GI and Van Duijn P (1986) A non-radioactive *in situ* hybridization method based on mercuratednucleic acid probes and sulfydryl hapten ligands. Nuc. Acids Res. **14**: 6471–6488.

Hopman AHN, Poddighe P, Smeets AWGB, Moesker O, Beck JLM,Vooijs GP and Ramaekers FCS (1989) Detection of numerical chromosome aberrations in bladder cancer by *in situ* hybridization. Am. J. Pathol. **135**: 1105–1117.

John H, Birnstiel M and Jones K (1969) RNA:DNA hybrids at the cytological level. Nature **223**: 582–587.

Keller GH, Huang DP and Marak MM (1989) Labeling of DNA probes with a photoactivatable hapten. Anal. Biochem. **177**: 392.

Kibbelaar RE, Van Kamp H, Dreef EJ, Wesseles JW, Beverstock GC, Raap K, Fibbe WE, den Ottolander GJ and Kluin PhM (1990) Detection of trisomy 8 in hematological disorders by *in situ* hybridization. Cytogenet. Cell Genet. in press.

Kievits T, Dauwerse JG, Wiegant J, Devilee P, Breuning MH, Cornelisse CJ, Van Ommen GJB and Pearson PL (1990) Rapid subchromosomal localization of cosmids by non-radioactive *in situ* hybridization. Cytogenet. Cell Genet. **53**: 134–136.

Kitigawa Y and Stollar BD (1982) Comparison of poly(A).poly(dT) and poly(I).poly(dC) as immunogens for the induction of anti-bodies to DNA:RNA hybrids. Molecular Immunol. **19**: 413–420.

Landegent JE, Jansen in de Wal N, Baan RA, Hoeijmakers JHJ and Van der Ploeg M (1984) Acetylaminofluorene-modified probes for the indirect hybridocytochemical detection of specific nucleic acid sequences. Exp. Cell Res. **153**: 61–72.

Landegent JE, Jansen in de Wal N, Van Ommen GJB, Baas F, De Vijlder JJM, Van Duijn P and Van der Ploeg M (1985) Chromosomal localization of a unique gene by non-radioactive *in situ* hybridization. Nature **317**: 175–177.

Landegent JE, Jansen in de Wal N, Dirks RW, Baas F and Van der Ploeg M (1987) Use of whole cosmid cloned genomic sequences for chromosomal localization by non-radioactive *in situ* hybridization. Hum. Genet. **77**: 366–370.

Langer PR, Waldrop AA and Ward DC (1981) Enzymatic synthesis of biotin labelled polynucleotides: novel nucleic acid affinity probes. Proc. Natl. Acad. Sci. USA **78**: 6633–6637.

Lawrence JB, Villnave CA and Singer RH (1988) Sensitive high resolution chromatin and chromosome mapping *in situ*: presence and orientation of two closely integrated copies of EBV in a lymphoma line. Cell **52**: 51–61.

Lawrence JB, Singer RH and Marselle LM (1989) Highly localised tracks of specific transcripts within interphase nuclei visualized by *in situ* hybridization. Cell **57**: 493–502.

Lawrence JB, Singer RH and McNeil JA (1990) Interphase and metaphase resolution of different distances within the human dystrophin gene. Science **249**: 928–931.

Lichter P, Tang CC, Call K, Hermanson G, Evans G, Housman D and Ward DC (1990) High-resolution mapping of human chromosome 11 by *in situ* hybridization with cosmid clones. Science **247**: 64.

Nederlof PM, Robinson D, Abuknesha R, Wiegant J, Hopman AHN, Tanke HJ and Raap AK (1989a) Three colour fluoresence *in situ* hybridization for the simultaneous detection of multiple nucleic acid sequences. Cytometry **10**: 20–27.

Nederlof PM, Van der Flier S, Raap AK, Tanke HJ, Van der Ploeg M, Kornips F and Geraedts JPM (1989b) Detection of chromosome aberrations in interphase tumour nuclei by non-radioactive *in situ* hybridization. Cancer Genet . Cytogenet. **42**: 87–98.

Nederlof PM, van der Flier S, Wiegant J, Raap AK, Tanke HJ, Ploem JS and Van der Ploeg M (1990) Multiple fluorescence *in situ* hybridization. Cytometry **11**: 126–131.

Pardue ML and Gall JG (1969) Molecular hybridization of radioactive DNA to the DNA of cytological preparations. Proc. Natl. Acad. Sci. USA **64**: 600–604.

Pinkel D, Straume T and Gray JW (1986) Cytogenetic analysis using quantitative, high sensitivity fluorescence hybridization. Proc. Natl. Acad. Sci. **85**: 2934–2938.

Raap AK, Marijnen JGJ and Van der Ploeg M (1984) Anti DNA.RNA sera. Specificity tests and application in quantitative *in situ* hybridization. Histochemistry **81**: 517–520.

Raap AK, Van der Ploeg M, Hopman AHN, Landegent JE and Van Duijn P (1987) Localization of DNA sequences by non-radioactive *in situ* hybridization. In: *Clinical Cytometry and Histometry*, pp. 221–226, Academic Press.

Raap AK, Hopman AHN and Van der Ploeg M (1989) Hapten labeling of nucleic acids probes for DNA *in situ* hybridization. In: *Techniques in Immunocytochemistry*, Vol IV. (Bullock G and Petruzs P eds.), pp. 167–198, Academic Press.

Raap AK, Dirks RW, Jiwa NM, Nederlof PM and Van der Ploeg M (1990) *In situ* hybridization with hapten-modified DNA probes. In: *Modern Pathology of AIDS and other Retroviral Infections*. (Racz P, Haase AT, Gluckman JC, eds), pp. 17–28, Karger, Basel.

Rudkin GT and Stollar BD (1977) High resolution detection of DNA-RNA hybrids *in situ* by indirect immunofluorescence. Nature **265**: 472–473.

Smit VTHBM, Wessels JW, Mollevanger P, Schrier PI, Raap AK, Beverstock GC and Cornelisse CJ (1990) Combined GTG-banding and non-radioactive *in situ* hybridization improves characterization of complex karyotypes. Cytogenet. Cell Genet. in press.

Schmitz H and Kampa D (1979) Amplified direct immunofluorescence (ADMI) for detection of Epstein-Barr virus nuclear antigen. J. Immunol. Meth. 26: 173–174.

Sverdlov ED, Monastyrskaya GS, Guskova LI, Levitan TL and Scheichenko VI and Budowski EI (1974) Modification of cytidine residues with a bisulfite-omethylhydroxylamine mixture. Biochim. Biophys. Acta **340**: 153–165.

Tanke HJ (1989) Does light microscopy have a future? J. Microsc.155: 405–418.

Tchen P, Fuchs RPP, Sage E and Leng M (1984) Chemically modified nucleic acids as immunodetectable probes in hybridization experiments. Proc. Natl. Acad. Sci. USA **81**: 3466–3470.

Teyssier JR (1989) The chromosomal analysis of human solid tumours: a triple challenge. Cancer Genet. Cytogenet. **37**: 103–125.

Trask B, Pinkel JD and Van den Engh G (1989) The proximity of DNA sequences in interphase cell nuclei is correlated to genomic distance and permits ordering of cosmids spanning 250 kilobase pairs. Genomics **5**: 710–717.

Van Prooijen-Knegt AC, Van Hoek JFM, Bauman JGJ, Van Duijn P, Wool IG and Van der Ploeg M (1982) *In situ* hybridization of DNA sequences in human metaphase chromosomes visualized by an indirect fluorescent immunocytochemical procedure. Exp. Cell Res. **141**: 397–407.

Vindelov LL, Christensen IJ, Jensen G and Nissen NI (1983) Limits of detection of nuclear DNA abnormalities by flow cytometric DNA analysis. Results obtained by a set of methods for sample storage, staining and internal standardization. Cytometry **3**: 332–339.

Viscidi RP, Conelly CJ and Yolken RH (1986) Novel chemical method for the preparation of nucleic acids for non-isotopic hybridization. J. Clinic. Microbiol. **23**: 311–317.

Wiegant J, Galjart N, Raap AK and d'Azzo (1992) The gene encoding human protective protein is on chromosome 20. Genomics: in press.

Index

MIX
Papier aus verantwortungsvollen Quellen
Paper from responsible sources
FSC® C105338

If you have any concerns about our products,
you can contact us on
ProductSafety@springernature.com

In case Publisher is established outside the EU,
the EU authorized representative is:
Springer Nature Customer Service Center GmbH
Europaplatz 3, 69115 Heidelberg, Germany

Printed by Libri Plureos GmbH
in Hamburg, Germany